# Environmental Ethics

## A Sourcebook
## for Educators

Bob Jickling, Heila Lotz-Sisitka, Lausanne
Olvitt, Rob O'Donoghue, Ingrid Schudel,
Dylan McGarry & Blair Niblett

SUN PRESS

*Environmental Ethics — A Sourcebook for Educators*

Published by African Sun Media under the SUN PReSS imprint

This publication was subjected to an independent double-blind peer evaluation by the publisher.

The authors and the publisher have made every effort to obtain permission for and acknowledge the use of copyrighted material. Refer all enquiries to the publisher.

Views reflected in this publication are not necessarily those of the publisher.

First edition 2021

ISBN 978-1-991201-28-7
ISBN 978-1-991201-29-4 (e-book)
https://doi.org/10.52779/9781991201294

Set in Lora 10.5/13

Typesetting and production by African Sun Media
Cover design: Heidie Aitken
Cover image: © Dylan McGarry

SUN PReSS is an imprint of African Sun Media. Scholarly, professional and reference works are published under this imprint in print and electronic formats.

This publication can be ordered from:
orders@africansunmedia.co.za
Takealot: bit.ly/2monsfl
Google Books: bit.ly/2k1Uilm
africansunmedia.store.it.si *(e-books)*
Amazon Kindle: amzn.to/2ktL.pkL

Visit africansunmedia.co.za for more information.

# Table of Contents

## CHAPTER 10 .................................................................................................. 221

*How elusive is ethical action?*

## CHAPTER 11 .................................................................................................. 241

*Implications for teaching and learning*

## AFTERWORD .................................................................................................. 265

*Contextualising environmental ethics in the contemporary global world*

## INDEX ........................................................................................................... 275

# ACKNOWLEDGEMENTS

Writing a book like this is inevitably a collaborative process. In many ways, it begins with our students who are often the first to try our new activities. It is their engagement, enthusiasm, and critical outlook that often drives our own commitment. This book is, in part, both for them and because of them. Thank you.

We would also like to thank other writers who contributed text that expands our horizons, reviewers who have helped to sharpen our thinking, and workshop participants who have offered guidance through their own engagement and reflections. We have been able to identify many of these groups and individuals below. We also thank those who helped but eluded our lists.

Finally, this book builds on past work, particularly the small book *Environmental Education, Ethics, and Action: A Workbook to get started.* We thank all of the participants who contributed to that volume, especially Dr Akpezi Ogbuigwe. We would also like to thank the United Nations Environment Program (UNEP) for its critical role in facilitating the book's completion and publication in 2006.

## Contributors

We are grateful to contributors who have added text to this book. First, we would like to thank Anthony Weston for his generosity in sharing his ideas about self-validating reduction and for his leadership in re-imagining environmental ethics. These sections of this book build on work originally developed by Anthony. Other contributors include Vivian Wood Alexander, art educator, Thunder Bay Art Gallery, Canada; Saransh Sugandh, film-maker, environmental educator based in Delhi; Xola Mali, South African blogger; Nikki Köhly; and Mirian Vilela, the Executive Director of the Earth Charter International Secretariat at the University for Peace in Costa Rica.

## Workshop contributions

Some of the significant venues for testing our work included: The World Congress on Education for Sustainable Development (ESD), Aichi-Nagoya, Japan; The Sustainable Campus Initiative Mentor Conference, Abu Dhabi; The Education as a Driver for Sustainable Development Goals Conference, Ahmedabad, India; The 8th World Environmental Education Congress, Gothenburg, Sweden; United Nations Environment Assembly 2, Nairobi, Kenya; The 2018 Conference of the North American Association for Environmental Education, Spokane, USA; and The 7th International Conference on Ethics Education, Porto, Portugal.

Individuals who facilitated these workshops and those who participated include:  Mirian Vilela, Arjen Wals, Waverli Maia M. Neuberger, Gayatri Raghwa, Kartikeya Sarabhai, Feruzan Mehta-Gwalior, Farida Vahedi, Betsan Martin, Khumansinh Shankarsinh Vaghela, Bhavesh Arvindbhai Trivedi, Saransh Sugandhi, Fatin Samara, Sinae Kim, Hyoeui Han, Kiran Chhokar, Sachi Ninomiya-Lim, Colin Bangay, Ranjana Saika, Ramjee Nagvajan, Jainy Shah, Tushar Dave, Irshad N. Theba, Alap Pandit, Sarah Dobson, Ms. Vidya Todankar, Kartikeya Sarabhai, Ulrica Stagell, Karen Jordan, Essi Aarnio-Linnanvuor, Maria Daskolia, Barbara Schäfli,  Auður Pálsdóttir, Karin Sporre, Patricia Armstrong, Rebecca Machin, Ellen Almers, Lydia Nicollet, Sophie Nicol, Marceline Collins-Figueroa, Kelsie Prabawa-Sear, Torbjörn Blanksvärd, Helene Grantz, Rosemary Black, Malin Gemzell, Deyne Meadow, Annica Corell, Katrine Dahl Madsen, Astrid Steele, Anna Görner, Andrea Welz, Annika Manni, David O. Kronlid, Ingela Bursjöö, Anna Heikkinen, Thao Phan Hoang Thu, Sue Elliott, Matsuda Takeshi,  Yoshiyuki Nagata, Kennert Danielsson, Marika Kose, Maria Letizia Montalbano, Nadia Lausselet, Aage Jensen, Louisa Chinyavu, Jared Onyari, Richard Kakeeto, Daniel Ehagi, Salaton Tome, Stephen Machua, Celestine Njuguna, Daniel Wepukhulu, Judy Braus, Ginger Potter, Peter Corcoran, Libby Bode, Katie Colleran, Nicki Dardinger, Richard Kool, Tara Muenz, Kelsey Vollmer.

## Financial contributions

We are grateful to the following funders for enabling the production of this book:
- the Symons Trust for Canadian Studies at Trent University, Canada, and
- the National Research Foundation of South Africa via the Research Chair in Global Change and Social Learning Systems (Grant Number: 98767) and a research project on Developing Ethico-Moral Agency through Environmental Learning Processes (Grant Number: TTK180423323615), both at Rhodes University, South Africa.

# AUTHOR BIOGRAPHIES

**Bob Jickling** is Professor Emeritus at Lakehead University (Canada). As an active practitioner, he taught courses in environmental philosophy; environmental, experiential, and outdoor education; and philosophy of education. Jickling was the founding editor of the *Canadian Journal of Environmental Education* and he co-chaired the 5th World Environmental Education Congress in Montreal. He has also received the North American Association of Environmental Education's Awards for Outstanding Contributions to: Research (2009) and Global Environmental Education (2001). His most recent book is *Wild Pedagogies: Touchstones for Re-Negotiating Education and the Environment in the Anthropocene*. As a long-time wilderness traveller, much of his inspiration is derived from the landscape of his home in Canada's Yukon. **ORCiD**: 0000-0001-6953-5976

**Heila Lotz-Sisitka** holds the South African Research Chairs Initiative (SARChI) Chair in Global Change and Social Learning Systems. Based in the Environmental Learning Research Centre at Rhodes University (South Africa), Lotz- Sisitka is a Professor in Education. Her research over 25 years has focussed broadly on education system development and transformative social learning for green, more socially just and sustainable economies and societies at local, regional, and international levels. Heila's research produces knowledge at the science-society interface, focusing on how learning leads development. This is critical for enabling a climate resilient development path in South Africa, and for facilitating access to new job opportunities opening up within the green economy. **ORCiD:** 0000-0002-5193-9881

**Lausanne Olvitt** is an Associate Professor of Education at Rhodes University (South Africa), with research affiliation to the Environmental Learning Research Centre. She has over 25 years' experience in the field of environmental education, starting as a teacher running a school's environmental club, then working in the environmental NGO sector and now in higher education. Lausanne currently teaches a range of courses at undergraduate and postgraduate levels and has been an active member of the Environmental Education Association of Southern Africa for many years, and Deputy Editor of its academic journal, the *Southern African Journal of Environmental Education*. Lausanne's research interests are mostly oriented around the 'fuzzy', affective dimensions of sustainability learning such as relationality, ethics, values, justice, care, and moral agency. She tries to keep this work grounded in local, practical sustainability practices in Makhanda where she lives with a collection of dogs, cats and vegetables. **ORCiD:** 0000-0003-4286-2875

**Rob O'Donoghue** is Professor Emeritus at the Environmental Learning Research Centre, Rhodes University. In his research on environmental education processes of learning-led change, he has given close attention to Indigenous knowledge practices, social theory and environmental learning in post-colonial curriculum and community contexts. Recent work with critical realism has been probing generative research and transformative social learning. This has informed his work with the Fundisa-for-Change teacher professional development programme and the Amanzi-for-Food initiative on local food production using rainwater harvesting to mitigate climate variability in the Eastern Cape. He recently initiated the Hand-Print CARE programme on ethics-led action learning in school subject disciplines with the Centre for Environmental Education, Ahmedabad, India. This developed as a collaborative initiative with colleagues in Germany, Mexico and India that included a concern for research ethics in evaluation research and ESD.

**Ingrid Schudel** is involved in teaching and research in the Environmental Learning Research Centre, Rhodes University in South Africa. Her portfolios include post-graduate teaching and supervision, in-service teacher training, and teaching for undergraduate pre-service teaching courses. Her PhD focused on environmental education in a teacher professional development programme and examined the emergence of active learning as transformative praxis in foundation phase classrooms. Currently her research focuses on 'Transformative Teaching and Learning in a Complex World' in the fields of environmental learning, curriculum, science education and teacher professional development. **ORCiD**: 0000-0001-6206-4681

**Dylan McGarry** is an educational sociologist and artist from Durban, South Africa. He is a Senior Researcher at the Environmental Learning Research Centre (ELRC) at the University currently known as Rhodes. As well as a co-director of the Global One Ocean Hub research network. Dylan is the co-founder of Empatheatre a public storytelling methodology and production company. Dylan is a passionate artist and story-teller. Dylan is also the co-founder of the Institute of Uncanny Justness, which re-imagines learning, activism and justice through suitably strange creative practices. He explores practice-based research into connective aesthetics, transgressive social learning, decolonisation, queer-eco pedagogy, immersive empathy and socio-ecological development in South Africa. His artwork and social praxis (which is closely related to his research) is particularly focused on empathy, and he primarily works with imagination, listening and intuition as actual sculptural materials in social settings to offer new ways to encourage personal, relational and collective agency. **ORCiD**: 0000-0001-5738-3813

**Blair Niblett** is an Associate Professor at Trent University's School of Education (Canada), where he designs hands-on experiences to illustrate theoretical concepts both in and out of the classroom across undergraduate, preservice, and graduate education classes. He is currently the graduate director for the PhD in Interdisciplinary Social Research, and co-editor of the *Canadian Journal of Environmental Education*. Blair's research explores the nexus of environmental education, social justice, and experiential learning. His most current work examines Canadian manifestations of the forest and nature school movement in early years learning programs. He lives in Toronto, Canada, the traditional territory of the Anishinaabe, Mississaugas and Haudenosaunee peoples, where Blair's European ancestors settled around 1890. He is happiest when exploring the land by bicycle, snowshoe, or paddleboard. **ORCiD:** 0000-0001-6487-7782

# CHAPTER 1

## INTRODUCTION

### Why this book? Why now?

The survival of many people and societies, and the biological support systems of the planet, are at risk. This is not something that has been dreamed up by radical environmentalists; rather, it is a mainstream view. In fact, this claim is taken directly from a document adopted by the 193-member United Nations General Assembly in September 2015.[1] There have also been continuing reports on climate peril from the International Governmental Panel on Climate Change (IPCC) and on biodiversity loss from the World Wide Fund for Nature (WWF).[2] Environmental issues have clearly become a concern for citizens of all countries, and these concerns have been growing as pressures on the Earth's systems increase.

While the United Nations (UN) member states acknowledge social-ecological risks, more and more people, young and old alike, are realising that environmental issues will not be resolved by technical means only. In times of change, uncertainty, and stress, interest in values and ethics increases. There is, indeed, no escaping from questions of value; our task is to engage with values carefully. This entails separating superficial questions – however we each might describe them – from important ones, and then carefully considering these important questions of value. When we make this move to carefully think about and act on important questions, we are doing ethics.

Talking about ethics can be intimidating. It does not have to be this way, and ultimately, it must not be this way. The aim of this book is to take ethics out of philosophy departments and put it in the streets, in villages, and on the Earth. The objective is to help make ethics a daily practice.[3] It should not be an exotic activity performed by heroes, saints, and experts that reside elsewhere – it is a matter for everyone. Exercising our ethical abilities should become part of our normal everyday activity.[4]

---

[1]   United Nations, "Transforming our world".
[2]   See for example: IPCC, *Special report: Global warming of* 1.5 C, 2018, accessed September 24, 2020, https://www.ipcc.ch/sr15/; and WWF (World Wide Fund for Nature), *Living planet report–2020*, accessed September 24, 2020, https://f.hubspotusercontent20.net/hubfs/4783129/LPR/PDFs/ENGLISH-FULL .pdf
[3]   See Andrew Sayer, *Realism and Social Science.*
[4]   Here we are indebted to John Ralston Saul for this idea and imagery that helps pull together and focus our intentions for this sourcebook: See John Saul Ralston, *On equilibrium.*

In fact, ethical questions cannot be avoided. They are present in assumptions guiding the choices we make and the things we do. We can aim to examine our assumptions and discover these questions, and consider them deliberately. However, if we do not, unconsidered ethical positions will still be present in what we do. They will be implicit in our actions, but unnoticed. One aim for this book is to make ethics, as an area of inquiry, more accessible and user friendly.

There are many who recognise the importance of ethics, but we are still a long way from making ethics an everyday activity. Despite growing interest, ethics remains on the periphery of mainstream education. As long as this remains the case, the implicit message delivered to students, parents, teachers, schools, and citizens is that ethical practice is not particularly important.

*Environmental Ethics: A Sourcebook for Educators* aims to provide ways for educators to get started in everyday ways. We believe this book is timely. Global efforts to care for the planet often remind us of our need for ethics. People around the world want better relationships between themselves, within communities, between communities, and between nations. They know that this includes relationships between humans and the more-than-human world, or, for others, between humans and the rest of Creation.[5] In using the term "more-than-human world", we suggest that exploring new relationships with Earth not only benefits human beings and their needs (although we recognise how important these are), but also the needs and well-being of forests, fields, rivers, animals, creatures in the sea, and the atmosphere.

Many of the world's environmental problems are not resolved because ethics questions are not asked. Ethics, and all the challenges that go with it, can initiate discussions about living well, about justice, equity, and common interests. We provide some beginnings for these discussions openings, starting points, challenges, and even hope.

Finally, on terminology, we are interested in the larger concepts of education and ethics, and their possible relationships within environmental education. Boundaries are not clear and language usage shifts between environmental education and education, and between environmental ethics and ethics. When we talk about environmental education and environmental ethics we bring ourselves home with more

---

[5] We have used the term "more-than-human" coined by David Abram and made popular in some circles through his book, *The Spell of the Sensuous: Perception and Language in a More-Than-Human World.* This is a conscious effort by Abram, and us, to disrupt the human-centeredness in the term "non-human." In this spirit we have also begun to use the phrase "the rest of Creation," at first simply as another alternative to "non-human," but later as a gesture of inclusiveness. Here we think of aboriginal persons who often talk about a Creator, other faith communities, and those that believe in evolution as a process of Creation. For those who do not feel comfortable with these terms, we invite you to find, or invent, your own alternatives.

narrow conceptions that reflect interests within our field. When we shift to talk about education and ethics we are probing nuances derived from the broader concepts.

## A sourcebook

In imagining this project, we had a common vision – to improve our teaching, and to encourage students and teachers to consider ethics everyday. As environmental educators, we share a deep concern for the human and the more-than-human world. We are also interested in how humans can live in different ways – ways that are not destructive to other humans, and to the more-than-human world. Achieving our vision begins with finding ways to help willing educators get started.

The activities featured in this publication have been developed and tested over many years. The authors have used them in undergraduate and graduate courses and in workshops in many parts of the world. All participants were encouraged to provide feedback based on their practical experience to help make the book better. The authors were committed to making this book as useful as possible.

The primary audiences for *Environmental Ethics: A Sourcebook for Educators* are schoolteachers and other non-formal and informal educators, teacher trainers, college instructors, university professors, and other professionals with responsibilities for professional development in education. It is also aimed at those passionate environmental educators who seek ways to take their teaching more deeply into questions at the heart of unsustainable patterns of living. We encourage readers to adapt the ideas by incorporating local examples, stories, and case studies to increase resonance in different social-ecological settings.

At the very heart of this book is its commitment to helping educators get started. However, from the beginning we realised that the scope of environmental ethics was far greater than the range of activities used in our earlier book, *Environmental Education, Ethics, and Action: A Workbook to Get Started.*[6] This current book has retained the pedagogical spirit and commitment to providing openings and access to environmental ethics for educators. However, it has a much greater reach.

In selecting our priorities, we have not tried to survey the field of environmental ethics. Rather, we have looked at the field and identified some turns in thinking that pose interesting questions and new possibilities for ethics, and for educators. So, while this book does reach across the field of environmental ethics, it touches down in places that we think are exciting and educationally productive.

---

[6] See Bob Jickling, Heila Lotz-Sisitka, Rob O'Donoghue and Akpezi Ogbuigwe, *Environmental Education, Ethics, and Action.*

We also think that there are some educators who will want to explore environmental ethics more deeply. With this in mind, we have provided a short essay about the theoretical thinking grounding our activities at the end of each chapter. This will be, in some ways, explanatory, but it will also provide an opening to explore interesting theoretical places. The essays also identify notable primary sources for those who wish to read further.

While we want this book to be *inviting*, we have not shied away from harsh and difficult realities. Power, injustice, exploitation, greed, and degradation are some of the themes running through our stories.[7] We encourage readers to make the invisible values that shape our societies more visible. We also encourage readers to re-imagine the world in creative ways, to find alternatives, to seek answers, and to take action.

Indian activist and writer Arundhati Roy, in her book, *The Algebra of Infinite Justice*,[8] reminds us that there is beauty yet in this brutal, damaged world of ours, and she encourages us to seek out the beauty, to nurture it and to love it. Her words encourage us to think about the future as open, and we recognise that "all manner of unimagined possibilities surround us". We think that an exploration of ethics is an important way to create an *open*, imaginative, just, and beautiful future.

In the end, however, this book is a beginning. No book can cover all ethical possibilities and one set of authors cannot speak for everyone. It is a starting point that invites people from all over the world to seek examples in their own communities, and from different cultural perspectives. It is the simple perspectives of one's local environmental values that can eventually bring out differentiated yet globally unified action towards the conservation of the environment. We encourage readers to continue the work begun here – to continue writing stories and imagining activities. We would like to hear about your innovations. This is vital.

## Using this book

The first aim of this book is to fill a void in providing practical material for teachers and teacher educators so as to begin incorporating ethics as an everyday activity into their teaching. All the chapters in the book introduce different pedagogical approaches for bringing ethics to the fore in environmental education. The activities in these chapters aim to be accessible, inviting, usable, and creative. In many instances there are also

---

[7] Here we take full account of our own ideologies. For example, through the examples we have included in the book, we signal dissatisfaction with consumerism, transnational corporations, and policies that work against social justice and a concern for the environment. We thus state our position clearly here, to avoid our own "absences," and invite readers to critically engage with the themes opened up by our choice of examples and language.

[8] See Arundhati Roy, *The Algebra of Infinite Justice*.

extension activities for those who wish to dive a little more deeply into the chapters' themes.

## Structure of the chapters

The book is also designed for accessibility. Each chapter begins with a preview of its contents. This includes a short statement describing the pedagogical intent, followed by a listing of activities in the chapter, sometimes organised around one or more subthemes. Each chapter then closes with a brief essay about some of the key theoretical resources that underpin the activities. We will say a little more about the theoretical resources later. First, though, we will talk a little more about the activities.

## Activities

The activities are all structured around a theme or series of themes and subthemes. They all begin with a short introduction to each theme that contextualises the inherent issues and identifies some pertinent observations and questions. The introductions are followed by concrete activities. Some activity sections also include extension activities. Some extensions allow teachers and learners to adapt the activities to their own schools or communities. Others are designed to be conducted over longer periods of time. Still, others invite learners to dig more deeply into an issue – practically and/or theoretically.

Key feature of activities in this sourcebook have been designed to help teachers "get started". There is no single correct way to take the first step. First, educators looking for particular kinds of activities – indoor, outdoor, group, individual, active, reflective, creative, critical, and complex – will find a broad range of possibilities to choose from. In this way, a practitioner can begin with a pedagogical approach that aligns with their teaching styles. This can be a steppingstone into more broadly engaging with environmental ethics.

Second, readers who want activities linked to particular theoretical orientations will find a range of possibilities. For example, some practitioners may have had some experience in animal welfare issues, or Deep Ecology and wish to begin with something that they know a little about.

Third, others who wish to engage themselves and their students in a more comprehensive survey of the field of environmental ethics, particularly from an education perspective, will find this book useful, too. In these instances, practitioners may wish to take a more-or-less sequential approach to engaging with this book.

As you get started, remember that there are many different ways to teach ethics. This book introduces activities, such as analysing images in popular media, discussing complex issues, contemplating local and international scenarios, reflecting upon daily actions, examining the stories that are told around us, and creatively imagining new possibilities.

Remember, too, that teaching about ethics is not a value-neutral activity. You will choose activities and teaching strategies just as we have done in writing this book, and no one can cover all ethical possibilities. You may develop some new activities that reflect your own interests and your approach to teaching. While teaching ethics is not value-neutral, it is not about imposing ideologies or seeking consensus either. Our challenge is to encourage processes of thoughtful deliberation and creative actions. These thoughts bring us to considering the theoretical background underlying the chapters.

## Handouts

Some activities include pages that are designed along the lines of traditional handouts. These are discrete pages that are intended to assist learners in completing the activities by summarising some of the analytical tools to be used, or to provide cultural examples that can enhance learners' understanding of the activities, or examples that can be analysed.

## Some thoughts about theory

The final section in each chapter is called "some thoughts about theory". This reading is not a prerequisite to using the activities. However, it is where educators will go if they want to understand more about ideas underlining activities in each chapter. For example, some readers might notice that our descriptions of this sourcebook – including its aims and activities – appear less structured and less prescriptive than they expected in an ethics book. In many cases this is because environmental ethics has been in a period of experimentation. Earlier practices in ethics have been challenged and insightful new questions are being asked. This final section provides a brief gateway into the more theoretical thinking behind the activities. It also provides links to some key writing by the major thinkers relevant to the themes in each chapter, for those that want to dig evermore deeply into this thinking.

These shifting perspectives and their implications for ethics in the 21st century are fittingly discussed in more detail at the end of this chapter in its "some thoughts about theory". Essays of a similar style appear at the ends of all chapters.

## Theory as stories

As it turns out, rising scepticism and complexities of ethically considering both humans and more-than-humans, has opened more avenues for doing ethics. It has also given us more questions. For example, we have now been encouraged to ask whether ethical compliance must rest with the concept of duty. Is it our duty to behave in ways deemed ethical? Or is there another way to go forward? Sometimes we know what we should do but don't take action. So, in thinking about this, we can ask: Where does the moral impulse to act come from? How can we nurture this impulse? Taking ethics in another direction, we can ask: Must ethics be strictly rational? Or should emotions be involved? Even, do we need a new kind of rationality? Also, the relationship between ethics and what we can learn has come into question. This expanded view of ethics in the 21st century is the subject of the "some thoughts about theory" section at the end of this first chapter.

Before continuing, there are a couple of important points to raise. First, given such an array of new possibilities within environmental ethics, we are not suggesting that anything goes. Holding some doubt about objective truth does not mean that we hold the particular relativistic view that everyone's truth, or theory, is of equal merit.[9] We believe that belief-systems that are incommensurate with each other can and often must be compared. We must weigh the merits of these systems and make judgements in the process of taking actions. We will discuss this further later in this chapter, and again in Chapter 10. We also believe that studying environmental ethics must not freeze our inclinations and abilities to act. Sometimes we must decide what constitutes an adequate basis for action. This, too, is discussed later in this chapter.

Second, a question that goes to the heart of our framing of this sourcebook: If theories are imperfect, can they still be useful? We think they can, especially if they can be thought of a little differently. If we agree that there is unlikely going to be a single theory that will define environmental ethics – a theory that will work for everyone – then what are we left with? We hold that there is still a lot of wisdom in past theorising. There will be more wisdom in the theorising yet to come. One way to appreciate this wisdom is to think about theories as stories, and a good story is one that has the capacity to do work. In our case this means good ethical work. Thus, much of the discussion and many of the activities in this book are framed as stories.

A commonality amongst activities in this book is that they encourage educators to examine stories that shape the world – stories found in newspapers, on television, on the streets, in the fields, and forests. These activities also encourage group discussion about complex issues, contemplating local and international scenarios, reflecting upon daily

---

[9]  The concept of relativism is too complex to discuss in detail in this sourcebook. However, for now we do want to be clear that we do not hold to the particular view that if objective truth is elusive, it follows that every person's truth is of equal value. We reject the sometimes-associated notion that there are facts and "alternative facts".

actions and the stories that are told around us, and creatively imagining new possibilities. We encourage educators and learners to look at things people do, in their daily lives, and in citizen actions. Mostly, we encourage educators and learners to tell stories – stories that are ethically inspiring, and stories that work.

Given that so much of this book is storied, we have expanded the discussion about stories in Chapter 2 of this sourcebook.

## Summary of chapters from a pedagogical perspective

Ethics, as a complex and nuanced matrix of possibilities, touches on many concepts important to education and teaching. As such, each chapter in this book begins with a pedagogical orientation. All the activities require a critical outlook on pedagogy.

In this Chapter, 1, we introduce the possibility of seeing theories as stories that reflect wisdom and have the capacity to do work. Chapter 2 develops this framing theory in storied terms – as a story at work. This chapter also introduces narrative voices and storytelling as a pedagogical approach.

Chapter 3 marks a beginning for broader social-ecological engagement and critique. This is a chapter designed to probe some of the baseline assumptions at play within our own cultures. Here the text and pedagogical activities aim to reveal the values that are buried so deeply in cultural artefacts – such as advertising, news reporting, and curricula – as to be nearly invisible. The pedagogies developed in Chapter 3 encourage learners to be evermore critical of the language, metaphors, and cultural practices that tend to preserve these values and reinforce the status quo.

Chapter 4 introduces examples of real-life environmental conflicts where contesting values result in difficult quandaries. While there are seldom clear answers in such quandaries, actions are often required, and judgements must be weighed and made. Pedagogically, learners are encouraged to explore stories with no easy answers. Many of these stories appeal to theorising around animal welfare issues. This chapter develops the idea that that even flawed theories can produce useful ethical work, particularly when represented as action-guided stories rather than theoretical truths.

Pedagogically, Chapter 5 links experiential learning and ethical engagement. The activities here rest on ideas that diverge from traditional approaches to environmental ethics. Rather that appealing to ethical duties, these activities seek to develop relationships with the world that can inspire more positive motivations for going forward – beautiful actions, and even joy.

While Chapter 4 introduces complexity in ethical decision-making, Chapter 6 expands on this. Pedagogically, the learning opportunities in this chapter are enriched through increasingly complex learning opportunities with attention given to specific cultural and historical contexts and feelings. They bring concrete realities to ethical deliberations in a real world.

Chapter 7 introduces a much different orientation to the role of ethics and how humans come to understand the world. Here activities challenge learners to examine relationships between ethics and what and how we can learn about the world. Here ethics – sometimes framed as etiquette – are seen to come first. Here the pedagogical implications are profound, as are emerging ethical possibilities.

Chapters 8, 9, and 10, each in its own way, engage pedagogies of social practice and learning. In each, ethics emerge from pedagogically-oriented activities, experiments, and real-life experiences.

In Chapter 8, learners are presented with examples where individuals can only fully understand the values they hold when they are challenged to act upon them in concrete situations. Or, in other words, learners are asked to consider how they may have learned something about their own values when they have been required to choose between different ways of handling a situation.

In important ways, Chapter 9 builds on and responds to Chapter 3. In the earlier chapter learners are encouraged to experiment with easily accessible analytical tools to critically examine values inherent in their social contexts – they deconstruct these socio-ecological spaces where they reside. However, it seems to us that deconstruction makes most sense when we then attempt to reconstruct something new, something better; and so, the theme of this chapter is re-imagining the future. Pedagogically, this work begins with the premise that learners need emotional, intellectual, and physical space to experiment with emerging ideas and to reimagine possibilities for creative alternative actions. Through environmental practice of trying out these new visions, space is created for ethics to arise.

Chapter 10 delves evermore deeply into the difficult and elusive nature of taking ethical action. Learning activities are designed to assist learners in navigating increasingly complex ethical landscapes with a view to finding paths to appropriate actions, especially in light of contesting claims.

Finally, Chapter 11 explicitly considers the complex relationships between environmental ethics and educational practices. An analytical tool is presented that can be used to examine complexities and challenges inherent in ethical theories, programmes, and codes, as well as the challenges involved in incorporating them into educational activities. Pedagogically, learners are encouraged to examine their own conceptions of ethics. Activities are then presented whereby they can – using The Earth Charter as an example – test these conceptions of ethics alongside their own conceptions of education and pedagogy. Finally, educators are encouraged to use this same analytical tool to examine pedagogical merits in examining other ethical theories and programmes.

Following these chapters we have provided a retrospective Afterword. Here we point to examples of worldwide environmental challenges and parallel moments of hope that occurred during the final years of writing this book. We do this to highlight the immediate and essential relevance of environmental ethics in the contemporary world – and the importance of education. We also include a reflective look at the emergent pedagogical strategies represented in this book.

All the chapters in the book introduce different pedagogical approaches for bringing ethics to the fore in environmental education. The broad pedagogical approaches, summarised in Figure 1 below, are interrelated in that they all help teachers to make ethics more visible, and to approach ethics teaching in ways that are engaged and linked to real situations or experiences in the world.

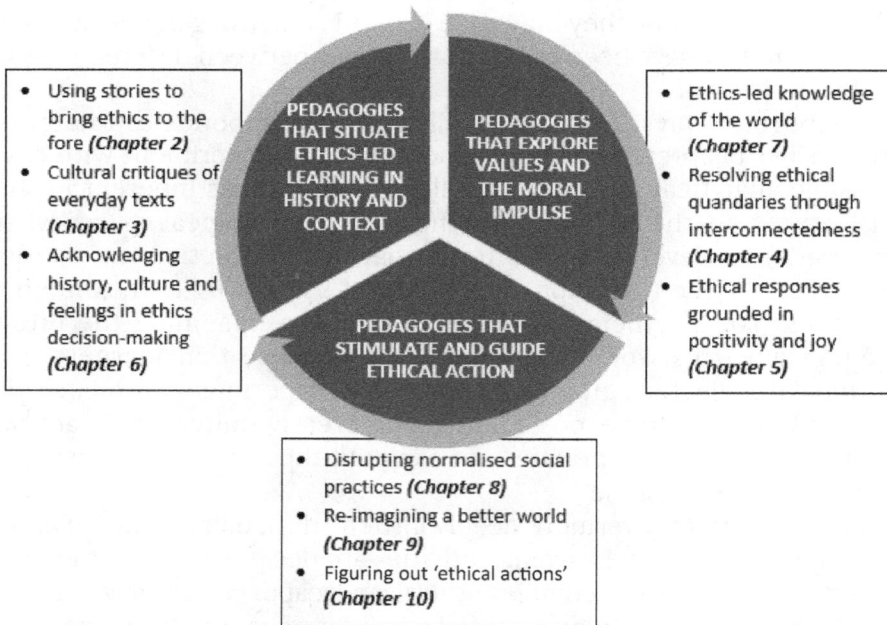

- Using stories to bring ethics to the fore *(Chapter 2)*
- Cultural critiques of everyday texts *(Chapter 3)*
- Acknowledging history, culture and feelings in ethics decision-making *(Chapter 6)*

**PEDAGOGIES THAT SITUATE ETHICS-LED LEARNING IN HISTORY AND CONTEXT**

**PEDAGOGIES THAT EXPLORE VALUES AND THE MORAL IMPULSE**

**PEDAGOGIES THAT STIMULATE AND GUIDE ETHICAL ACTION**

- Ethics-led knowledge of the world *(Chapter 7)*
- Resolving ethical quandaries through interconnectedness *(Chapter 4)*
- Ethical responses grounded in positivity and joy *(Chapter 5)*

- Disrupting normalised social practices *(Chapter 8)*
- Re-imagining a better world *(Chapter 9)*
- Figuring out 'ethical actions' *(Chapter 10)*

Figure 1.1    Pedagogical emphases across the chapters of this book

## Additional considerations

Having outlined the context, intent, and structure of this sourcebook there are a few more background topics that help to describe some of our underlying thinking in writing this sourcebook. In doing so, we have also identified some practical considerations that can support the practitioner in the work of going forward. As these are a little more theoretical, we have placed them in the section "some thoughts about theory" at the end of this chapter.

First, we have clarified our use of the words "values, ethics, and morals" and begun a conversation about how these concepts can be examined within educational practices. Some readers might find this useful now and others may want to get into the book right away and return to this later.

Second, we feel that throughout this book practitioners and learners will be faced with the task of weighing and judging options between contesting points of view. A section titled "weighting and judgement" provides some guidance. Sometimes, we find that we must act on incomplete information. The section on practical adequacy discusses what might be considered in everyday issues.

In the present era, an outsized amount of influence seems to be attributed to the United Nations' Sustainable Development Goals. They seem impossible to ignore. A final section discusses these goals and begins the task of weighing and judging their adequacy.

## SOME THOUGHTS ABOUT THEORY

### Some emerging trends in environmental ethics

We believe that environmental ethics has had a stimulating role – in the political arena and within the broader field of ethics. Forays into environmental ethics in the 1970s and 80s often resembled human ethics as attempts were made to extend these ethics to include consideration of other life forms. Theories about animal rights and animal liberation were prominent at the time. Typically, ethicists attempted to establish basic ethical principles or assumptions that could serve as a universal basis for ethical actions. The principles (sometimes referred to as axioms) were abstract ideas and putting them into practice relied on the established rationality of the times – a cognitive rationality stripped of emotional content. It quickly became evident that these and other approaches to environmental ethics gathered followers and momentum, and became guiding forces in environmental activism. A lot of ethical work was accomplished.

The second half of the 20th century also saw a rise in critical and postmodern theories, with their scepticism of utopic ideals, universal theories, and tendencies toward deconstruction through abstraction rather than grounded example. This rise awoke scepticism for ethics and ethical theorising that were thought to depend on universal narratives and singular ethical precepts that are unresponsive to context.[10]

---

[10] See Rosi Braidotti, *Posthuman Knowledge.*

Some folks working in environmental ethics were clearly influenced by the mood of the times and moved away from a quest for a single universal theory for the field, even if they did not fully embrace critical or postmodern theorising. Many remained committed to the importance of direct experiences in places and landscapes in the development of ethical thinking and practices. Some would not be constrained by the kinds of boundaries established within human ethics – especially when applied to relationships with the more-than-human world. For example, Anthony Weston[11] argues that our challenge is not to systematise environmental values, but to create the "space" for environmental values to evolve, particularly through what he calls environmental practices.

Our purpose here is not to describe all the twists and turns in the development of environmental ethics during the mid- and late- 20th century. Rather, we want to provide a flavour of the times and to show that environmental ethics was freed from some constraints of traditional ethics, and how this opens up interesting new possibilities. This, of course, enriches the field, but it also allows more entry points for the educator.

Some of the new entry points can be framed as different kinds of questions. We think that responding to these questions in practice will allow for new and different approaches – to start with different questions and to ask different things of environmental ethics and of education. The hope is that all educators will find chapters that will allow comfortable entry points best suited to their experiences. Taken together, the chapters illustrate the potential breadth in places where environmental ethics and education intersect. Consider:

- Why don't people care enough to act on the seemingly moral imperatives of our times?
- Where does the moral impulse – the urge to act – come from?
- Must ethics rely on duties?
- Can ethics be joyful?
- Can effective ethics be grounded in abstract principles?
- Or can they arise out of lived experience?

We are excited by the diversity of approaches and entry points for environmental ethics and the educator; however, we also urge caution. We are not promoting a kind of "anything goes" ethics; there are constraints. We remind readers that this book was published at a time of unprecedented ecological risk. Sustainability of myriad species and, indeed, humans is not guaranteed.

It is possible to have deep and thoughtful questioning and still come up with ideas that will result in unecological consequences. In cases like this, it is educationally appropriate to continue a respectful dialogue that seeks to further examine the ideas and to bring additional perspectives to the discussion. In the end we may respectfully disagree. However, this

---

[11]　Anthony Weston, "Before environmental ethics," 321-338.

process also requires humility and openness on the part of the educator. It is always possible that discussions can lead learners to come up with better ideas than their mentors.

Some ethicists might regard achieving mere "sustainability" a low benchmark for success when paired with more traditional ideas about living a good life or living well in a place. Here, environmental ethics can be about finding ways to be more considerate of more-than-humans – their pain and suffering, their rights, and even the inherent rights and integrity of ecosystems, bioregions, or watersheds. Framed another way, some will see progress in finding ways to build better lived-relationships with the more-than-human-world.

What is important throughout this section on environmental ethics, moving into the 21$^{st}$ century, is that the primary aim is not about seeking "ethical truths." Rather the task is to make progress towards eco-systemic sustainability, moral consideration of the more-than-human world, and achieving lived relationships with places of dwelling.

Rethinking ethics, as a move away from seeking truths in favour of making ethical progress, may also reflect what some have called a period of "theory fatigue". Some suggest that we are experiencing a post-theory mood.[12] Our proposal is to think of theories developed in environmental ethics as "stories", and, as stories, we are interested in their capacity to do work – or as we have said, to make progress. Our particular interpretation of stories has been inspired by Indigenous ways of knowing and is an important basis for framing this book. These ideas about stories are outlined in Chapter 2.

As educators our task is to give our students opportunity to get involved with the process of doing ethics – to find their own paths through emerging possibilities in environmental ethics. In our book, we see ethics as being situated in the world, linked to real, practical contexts in which they emerge. This makes ethics a contextual, and tension-laden affair, and an important ongoing process that must be worked out in the world with real people in real places.

Through these chapters, we provide multiple entry points to environmental ethics and a breadth of approaches, which can help both learners and educators become more aware of different ways of enabling ethics. Some of these approaches reflect an evolution of thinking in the field, others represent divergences, and sometimes the approaches presented are contesting. Perhaps the most important reason for our decision to frame the book this way is pedagogical. In standing against "ethical truths" we are raising important issues about pedagogy, weighting and judgement, and practical adequacy.

---

[12]  Rosi Braidotti, *Posthuman Knowledge*.

## Values, ethics, and morals in education

In this book, we explore creative ways of working with ethics in education. For many, ethics can be a difficult term, and sometimes a little scary. So, to begin with, we will clarify how we are using some key terms – values, ethics, and morals.[13]

Values are expressions of preferences and priorities – things that are more or less important to us. Values are the things we weigh as we make choices. Do I value this more than that? It is the broadest in our suite of key terms. As the broadest category, it encompasses the widest range of possibilities and topics. At one end of the spectrum are values as simple as consumer choices between brands, flavours, and colours, and rivalries between sports teams. At the other end, are more profound differences in the kinds of relationships we have in the world. These are the kinds of values that we extend beyond ourselves, towards concerns for others. In environmental ethics our concern is not limited to just human others.

Take for example the term "more-than-human world". This term was coined by David Abram in response to values embedded in the term "non-human."[14] His point was that when we talk about non-humans, we are implicitly setting up a hierarchy of values. That is, there are humans that are more important by virtue of being named, and there are non-humans that are less important by being treated as a simple aggregate. Through this new term he wanted to elevate the values recognised in the much larger portion of the world that wasn't human. If you are sceptical of Abram's logic, consider how, in Apartheid South Africa, the term "non-white" was used as a way of perpetuating white supremacy. It is this kind of implicit weighing of values that offended Abram.

Morals, or moral values, are a subset of the broader category of values. Whereas values include simple preferences, often of little consequence to most others, moral values concern those priorities that are more deeply shared and convey a sense of expectation or performance. As Anthony Weston says, moral values are "those values that give voice to the needs and legitimate expectations of others as well as ourselves."[15] Traditionally morals have been limited to human interactions, but environmental ethics challenges these limitations in many ways, including Abram's concern in the example above. So, morals are the values that we draw on to guide us in "living right" as we take action to enact our values. Sometimes these morals are individual and personal, and sometimes they are social expectations. Another related and perplexing concept is the idea of the moral impulse.

---

[13] Here we largely follow the lead of Anthony Weston in his book, A 21st Century Ethical Toolbox. There are other ways of framing these terms, but these are our convention for this text.

[14] David Abram, The Spell of the Sensuous.

[15] Anthony Weston, A 21st Century Ethical Toolbox, 12.

What is it that triggers our actions, and inspires us to act? This idea of the moral impulse will be explored a little later in this book (Chapter 8).

Those reading the previous paragraph on morals with a critical eye may have zeroed in on the word legitimate, as in "the legitimate expectations". What is legitimate? Whose idea of legitimate should count? Good questions, and they lead us to ethics. A little confusion sometimes arises when ethics and morals are used interchangeably. In this book, we make a distinction between them and use ethics to mean a process of inquiry that examines, for example, questions about legitimate expectations. Ethics as a field of study inquires about priorities, or moral values – how they are established, ways they can be shifted, and how they can be applied. It also includes inquiry about how actions can inform ethics.

Seen this way, ethics is not about "preaching", "indoctrinating", or "inducting" learners into "rules of behaviour" or "codes of conduct". Rather, ethics is an open-ended process with the potential to expose new challenges and generate new possibilities. It is a process of making choices that enable better ways of seeing and doing things. This does not mean that decisions and actions are never taken – we do act. It does mean, however, that ethical positions are open for discussion, re-examination, and revision.

Education would better serve the whole person if ethics were included in all subjects, thematic studies, and at all stages and levels. This is important work.

## Weighting and judgement

The quandaries in Chapter 4 and other complex ethical struggles reported in later chapters clearly show that in complex issues there are often contested claims between proponents of worthy values. Again, this is a reflection of diversity and complexity in ethical concerns.

For example, in Canada's Yukon, the Territorial Government conducted a controversial wolf kill in the 1990s. This meant that 80% of the wolves in a 20,000 square kilometre region were eliminated. The release for caribou and moose from hunting pressure by wolves can allow populations of moose and caribou to recover to more normally expected levels.

Some First Nation proponents argued for the necessity of the wolf kill on cultural grounds. For them, survival of cultural knowledge and practices was dependent upon the ability to hunt caribou. Since caribou numbers were in steep decline, aspects of cultural survival were threatened. The wolf kill was seen as a way to protect and preserve their cultures.

There was also an ecological perspective. One biologist sought to restore the ecological integrity of the region in question out of a deep affection for the landscape, complete with historical levels of caribou.

People operating from a third contesting position opposed the wolf kill. There were several threads to this opposition, but a key element rested on respect for wolves as a highly evolved social species. Within this position there was a belief that trying to solve the problem of declining caribou and moose numbers with technical solutions, such as massive aerial wolf killing, would not address root causes of the issues. For these opponents, the problem of declining moose and caribou populations was primarily caused by humans. They pointed to a combination of road building into critical habitats and over-hunting as major culprits. The feeling was that using modern technologies to kill wolves would do little to change relationships between humans and wolves where, in essence, wolves are seen by some to have no intrinsic value, and are essentially treated as vermin.

Moose and caribou were seen by some as commodities to be produced for eager consumers. This points to a fourth position that was omnipresent. There were many hunters who demanded that the government rid the land of wolves so their hunting opportunities would improve.

When comparing these positions, particularly the first three, it can be argued that they carry significant moral weight. Their proponents are not wrong to want to protect deeply held cultural values and traditions, or restoration of ecological integrity, or to enhance respect for wolves while placing the burden of ecological restoration on humans and curtailing the kinds of human actions that caused the problem in the first place.

In analysing the ensuing conundrum, there was a plurality of values at play. That is, there were multiple perspectives that carried ethical weight – they could not be dismissed as simply wrong. That said, the ensuing controversy bore no signs of ethical relativism – that is, in this context, these value positions were not treated with equal weighting. Instead, there were vigorous public arguments that extended over years. These arguments, sometimes intensely fierce, were essentially about weighing the merits of the claims and trying to work out which of them carried the greatest weight and hence, a basis for action.

The point here is that the existence of multiple perspectives, rather than operating from a universalised ethical system, does not necessarily imply, or lead to, the kind of relativism where all values are considered equal. Controversies, however messy, are ultimately about weighing claims and making judgements.

## Practical adequacy and the "common good"

Another way to look at this process of weighing and judging is as a search for practical adequacy. That is, we cannot possibly know everything, but sometimes we still must make decisions that are as thoughtful as possible, and with the best information available. This is a practical thing to do. Understanding ethical quandaries in real situations, using available stories, is a practical alternative to the failures of the kind of ethical relativism discussed in the previous section. But what is adequate?

History shows that humanity's knowledge, values, and beliefs may have appeared practical and adequate in a particular time and space to some. However, this practical adequacy has typically benefited a portion of a population but has not necessarily brought about good or practical adequacy for everyone. For example, the extractive policies of colonial and neo-colonial governments could be seen to be practically adequate for those who were benefitting from the extractive policies; yet they were wholly inadequate and exploitative when seen from the experiences of the colonised. Archbishop Desmond Tutu (a peace activist, and perhaps an ethicist in his own right) offers a pithy story of the limitations of practical adequacy: "If you are neutral in a situation of injustice, you have chosen the side of the oppressor. If an elephant has its foot on the tail of a mouse, and you say you are neutral, the mouse will not appreciate your neutrality."[16] For Tutu, the observer's neutrality may be practically adequate based on a distant position, but increasing proximity would make it difficult to reasonably ignore the sense of threat that the mouse experiences.

One way of judging adequacy is through a reference to the "common good". Here, though, we must also have a sense of meaning for this term. Some of the hard work on the common good, and its meaning, has been done by Roy Bhaskar. For him, the common good is related to flourishing, where the flourishing of one is related to the flourishing of all.[17] From this, we can begin to seek guidance in making judgements about practical adequacy or truth.

To make thoughtful judgements about practically-adequate responses to issues requires thinking about alternatives, or desired future states. Here we can think in terms of:

- How the recommended social processes would work in practice.
- Asking critical and difficult questions that probe these imagined possibilities with the utmost care; these questions include commitment to probe the feasibility of the possible alternatives.
- Conducting thought experiments to try out possibilities, to re-imagine different routes forward.
- Always scrutinising critical standpoints that challenge our taken for granted assumptions.

This book offers many activities that can help us with some of this work in educational settings. In this way, we can begin to develop ethics activities and practices that will be carefully constituted, and that are also feasible and practically adequate. Sensitivity to context is not equivalent to the notion of ethical relativism. It is possible to engage with ethics questions as stories, decision-making processes, and practices that are

---

[16] As cited in Gary Young, "The secrets of a peacemaker".
[17] See Roy Bhaskar, *Dialectic: The pulse of freedom*.

situated in and shaped by specific contexts. At the same time, we can also recognise that these decision-making processes form part of a wider sense of a common good. By common good, we are referring to a good that considers the needs of all humans and the more-than-human world, rather than "one" answer or one universalising ethic for all.

We started this book by talking about stories and their capacity to do ethical work. This is a little different from the exercise of developing ethical theories that seek to identify basic ethical principles and then to develop frameworks for universally applying these principles in a wide range of contexts. At this time, such a perfectly usable theory is not available.

In the meantime, we ask, despite flaws and irreconcilable conundrums, do these stories throughout this book still carry sufficient weight to allow us to do ethical work? We must make judgements in everyday issues and these ethical stories help. Short of providing universal truths that are frustratingly difficult to apply in all situations, we can still ask, do they have practical adequacy? That is, can they provide guidance when a practical judgements and actions are required?

The point here is that acknowledging the existence of multiple perspectives, rather than operating from a universalised ethical system, does not necessarily imply, or lead to, a kind of ethical relativism where all values are considered equal. Controversies, however messy, are ultimately about weighing claims and making judgements; this is a practical alternative to the failures of the kinds of relativism discussed in this book.

## Sustainable development goals, ethics, and education

Keen observers will notice that during the latter part of the 20th century and into the 21st century UNESCO has mounted an assertive campaign to promote education for sustainable development and the United Nations' 17 sustainable development goals. As such, these goals are an important issue of our times and warrant both consideration and scrutiny.

In September 2015, the 193-member United Nations General Assembly adopted the 2030 Agenda for Sustainable Development together with 17 Goals for implementation. This was hailed by the Secretary-General, Ban Ki-moon, as a "promise by leaders to all people everywhere. It is an agenda for people to end poverty in all its forms – an agenda for the planet, our common home."[18] Perhaps this resolution expresses a realisation that globalisation is essentially about an economically driven vision for the world, and that an approach which over-emphasises economics is failing humanity. So far, globalisation has not helped to resolve many of the world's most pressing problems. As Nelson Mandela said, years earlier, "Massive poverty and obscene inequality are such terrible scourges of our

---

[18]   UN News, "UN adopts new Global Goals".

times."[19] Perhaps, during this time, questions of ethics have been left in the background.

Interestingly, and immediately following the 2015 General Assembly, the Permanent Mission of Panama convened a panel discussion in January 2016. This panel was organised in collaboration with the Permanent Missions of Argentina, Costa Rica, Finland, Germany, Kazakhstan, Palau, and Qatar and was titled "Ethics and the Implementation of the Sustainable Development Goals (SDGs)". By hosting this panel discussion, the organisers made explicit what had always been implied: The Sustainable Development Goals rest on an ethical foundation. Or, put another way, these goals are ethics-led.

To underscore this ethical foundation, the UN Deputy Secretary-General Jan Eliasson noted that the 2030 Agenda "is grounded in, and guided by, the fundamental values and principles laid down in the UN Charter, in particular its preamble. It is also inspired by the Universal Declaration of Human Rights." He went on to add that the "fundamental principles that underpin the new goals are interdependence, universality and solidarity" and that this is its "underlying moral code," its "profound ethical foundation."[20]

While the Sustainable Development Goals can be said to be ethics-led, many of the goals are clearly political in implementation, that is, they point to things governments must do to meet the goals of the resolution. Interestingly, Goal 4 calls for quality education. As always, when talking about education and politics there is a tension between educational and political aims. Can, for example, quality education be limited to the implementation of political aims? And when we talk about education and ethics there is also a tension. For example, is it educational to reduce ethics to the implementation of a moral code? Are there other approaches to ethics besides developing and implementing moral codes? There are, of course, critiques of education for sustainable development agenda available in the literature that can provide some insight into these questions.[21]

When considering these questions about education, politics, and ethics, a former head of UNEP's Environmental Education and Training Unit remarked (slightly paraphrased), "You can say what you like about what might be missing in the SDGs, but it was an incredible job getting all of those countries to agree on these goals. It is a collective step in the right direction." This is an extraordinarily interesting and telling statement.

---

[19] Nelson Mandela, "Make History. Make Poverty History".
[20] United Nations Secretary-General, "Deputy Secretary-General remarks".
[21] See for example, Bob Jickling and Arjen Wals, "Globalization and environmental education: Looking beyond sustainability and sustainable development," 1-21; Helen Kopnina, "Education for sustainable development (ESD)," 699-717; Helen Kopnina, "Revisiting Education for Sustainable Development (ESD)," 73-83; and Sam Adelman, "The sustainable development goals," 15-40.

First, he is correct; the collective agreement does set many important priorities. Within these content areas there is much educational work to do. However, implicit in the nature of the UN Resolution's compromise, and the implication that there will be more steps to follow, suggest more educational questions. For example, what must we do educationally to ensure that the people are prepared to imagine what these next steps might be – beyond sustainable development, and the sustainable development goals?

The foregoing discussion is meant to introduce the ethics-led nature of important international initiatives, but it is also meant to introduce the reader to questions to ponder while travelling through this book. What aims do you hold for education? How can approaches to ethics be expanded, beyond just thinking about moral codes? How should educators approach ethics in general, and environmental ethics in particular? These questions are implicit throughout the book and are made explicit at intervals. We return to these questions in the closing sections of this book (Chapter 11).

While the foregoing are good questions to ask about relationships between the sustainable development goals and education, there are important questions to ask about the content of the goals themselves. At first glance they may have a lot of appeal to social and justice-minded educators and learners. However, it can also be good to look closely to determine what other values might be implied – or be implicit – by virtue of the way that the goals are written. Hand in hand with this analysis, it can also be important to look closely for what may be omitted. These implicit values and/or omissions are sometimes referred to as a hidden curriculum. Again, we can find starting points for these analyses in the literature.[22]

Fortunately, reading the literature is not the only way to begin investigating the hidden curriculum. Chapter 3 is designed to help learners recognize the kinds of values that, while sometimes hidden, shape public documents. In Activity 3.2: (A Curriculum Critique) there is an extension activity that specifically invites learners to conduct their own analysis of the sustainable development goals.

## References

Abram, David. *The Spell of the Sensuous: Perception and Language in a More-Than-Human World.* New York: Pantheon Books, 1996.

Adelman, Sam. "The sustainable development goals, anthropocentrism and neoliberalism," in *Sustainable Development Goals: Law, Theory and Implementation.* Edited by D. French and L. Kotzé. Cheltenham, Glos:

---

[22]  See for example, Elliot Eisner, "The Three Curricula All Schools Teach," 87-108.

Edward Elgar, 2018, 15-40. https://doi.org/10.4337/9781786438768
.00008

Bhaskar, Roy. *Dialectic: The pulse of freedom.* London: Verso, 1993.

Braidotti, Rosi. *Posthuman Knowledge.* Cambridge: Polity Press, 2019.

Eisner, Elliot. "The Three Curricula All Schools Teach", in *The Educational Imagination.* New York: MacMillan, 1985, 87-108.

Jickling, Bob, Heila Lotz-Sisitka, Rob O'Donoghue, and Akpezi Ogbuigwe. *Environmental Education, Ethics, and Action: A Workbook to Get Started.* Nairobi: UNEP, 2006.

Jickling, Bob and Arjen E.J. Wals. "Globalization and environmental education: Looking beyond sustainability and sustainable development." *Journal of Curriculum Studies* 40, no. 1 (2008): 1-21. https://doi.org/10.1080/00220270701684667

Kopnina, Helen. "Education for sustainable development (ESD): the turn away from 'environment' in environmental education?" *Environmental Education Research* 18, no. 5, (2012): 699-717. https://doi.org/10.1080/13504622.2012.658028

Kopnina, Helen. "Revisiting Education for Sustainable Development (ESD): Examining Anthropocentric Bias Through the Transition of Environmental Education to ESD." *Sustainable Development* 22, no. 2 (2014): 73-83. https://doi.org/10.1002/sd.529

Mandela, Nelson. "Make History. Make Poverty History." Speech, The Campaign to Make Poverty History, Trafalgar Square, London, United Kingdom, February 3, 2005. https://doi.org/10.9774/GLEAF.4700.2005.su.00015

Roy, Arundhati. *The Algebra of Infinite Justice.* London: Harper Collins, 2002.

Saul, John R. *On equilibrium.* Toronto: Penguin, 2001.

Sayer, Andrew. *Realism and Social Science.* London: SAGE, 2000. https://doi.org/10.4135/9781446218730

United Nations. "Transforming our world: the 2030 Agenda for Sustainable Development," September 25, 2015, accessed June 14, 2019, https://sustainabledevelopment.un.org/post2015/transformingourworld

United Nations Secretary-General. "Deputy Secretary-General remarks at ethics for development side event at the the UN General Assembly [as delivered]," January 13, 2016, accessed June 14, 2019, http://www.un.org/sg/dsg/statements/index.asp?nid=694

United Nations Deputy Secretary-General. "Statements," downloaded May 13, 2016, http://www.un.org/sg/dsg/statements/index.asp?nid=694

United Nation News. "UN adopts new Global Goals, charting sustainable development for people and planet by 2030," 25 September 2015, accessed June 14, 2019, http://www.un.org/apps/news/story.asp?NewsID=51968#.VwH2U84mXCO

Weston, Anthony. "Before environmental ethics." *Environmental Ethics* 14, no. 4 (1992): 321-338. https://doi.org/10.5840/enviroethics19921444

Weston, Anthony. A 21st Century Ethical Toolbox. New York: Oxford
    University Press, 2001.
Young, Gary. "The secrets of a peacemaker," The Guardian, May 23, 2009,
    accessed June 14, 2019, https://www.theguardian.com/books/2009
    /may/23/interview-desmond-tutu

# CHAPTER 2

## WHAT WORK CAN OUR STORIES DO?

**Pedagogical intent:** This chapter shows that stories are an age-old approach to sharing ethics perspectives in society. It encourages teachers to use stories that illustrate ethical quandaries, guide actions, and bring ethics questions and insights to the fore.

**Activity:**                    2.1: Storying environmental ethics

**Main theoretical resources:** The theoretical reflections begin with First Nation stories and thoughts on the nature and value of stories. Storytelling is then traced through 19th century literature, contemporary television, and more recent environmental ethicists, particularly Aldo Leopold and Jim Cheney.

## A story at work

Current thinking in environmental ethics is beginning to illustrate understanding of the importance of stories. For many, ethics without a storied context lacks nuance, history, and place. Thus, the narrative voice is being rediscovered by educators and ethicists seeking more authentic ways of speaking, doing, and living their work. Aldo Leopold's *Sand County Almanac*,[1] and Henry David Thoreau's *Walden*[2] are richly storied accounts of encounters with landscapes, and harbingers of a "modern" trend.

More recently, many of Vandana Shiva's books are both storied and ethics-centred. For example her book *Making Peace with the Earth* argues against "war with the earth" or ongoing destruction of the earth and her resources.[3] In her book *Earth Democracy: Justice, Sustainability and Peace*,[4] she calls for a radical shift in the values that govern democracies. Using stories of people's lives and experiences from streets and farms around the world, she lays bare the consequences of unchecked capitalism via descriptions of critical issues such as genetic food engineering, natural resources privatisation, culture theft, and violence against women.

Wangari Maathai's story of the Green Belt Movement in Kenya and Africa is a practical ethical expression of how things can change towards earth democracy, peace, justice, and sustainability.[5]

Aboriginal peoples around the world have always understood the importance of stories. It is implicit in oral cultures. Carol Geddes, a Tlingit

---

[1]    See Aldo Leopold. A *Sand County Almanac.*
[2]    Henry David Thoreau, *Walden.*
[3]    See Vandana Shiva, *Making peace with the earth.*
[4]    Vandana Shiva, *Earth democracy: Justice, sustainability and peace.*
[5]    Wangari Maathai, *The Green Belt Movement.*

woman from Canada, summarised this succinctly when she said that in her culture there would not be a subject called ethics; it would all be part of the story. In other words, ethics cannot be separated from tales told, knowledge learned, daily practices, and lives lived.[6]

Stories permeate this book. As Robert Bringhurst says, stories tend to "sprawl all over each other".[7] Some activities will be storied, and stories will add a layer to the theorising pieces that end each chapter.

Still, for those of us who have walked solidly on the ground of rationalist traditions, and those of us who are mainly exposed to these traditions in and through our modern-day education systems, it can be one thing to have an inkling about storied possibilities – as mirrors, relationships, nuanced experiences, and lived lives – it is another to stand on the fertile earth of story. With this in mind, our first activity is centered on a story at work.

To be grounded, the central story in this activity needs to be contextualised. Tell it outside. Find a quiet place suitable for reflection, and even better, a place that is also inspiring. Though, an outside place that appears bleak could also work. Because the story takes place in winter, we like to take up this activity in that same season. As learners sit and listen, context is added if there is a chill in the air.

## LEARNING ACTIVITY 2.1: STORYING ENVIRONMENTAL ETHICS

Prepare your group to go outside, suitably attired for the climate and weather. While appropriate clothing can generally be expected – for where you live – people are not always prepared to sit down while outside. We find that it is handy to have on hand something to sit on that can protect the learners and their clothing from the ground. This can be as simple as a group set of plastic bags or pieces of cardboard or, more elaborately, small mats or cushions.

1. Lead your group to the chosen location, then have everyone sit in a circle. When settled, calm the group, encourage them to listen closely. You could remind them that sometimes in oral cultures it is not always the speaker's job to speak loudly; but rather, it is the task of the listener to be more attentive. When everyone is settled, read aloud the following story told by Louise Profeit-Leblanc:[8]

   Let me now share with you a story which takes place in the Blackstone River district in northern Yukon, up the Dempster Highway. This is the

---

[6]  Carol Geddes, in a panel discussion, "What is a good way to teach children and young adults to respect the land?", 32-48.

[7]  Robert Bringhurst, "The tree of meaning," 16.

[8]  Louise Profeit-Leblanc, "Transferring wisdom through storytelling," 15-17.

homeland of the Takudh Gwichin. There are three groupings of Gwichin, the Tatlit Gwichin who inhabit the Arctic Red River and Fort McPherson region of the NWT, the Takudh Gwichin from the Blackstone and the Vuntat Gwichin from Old Crow. Apparently, years ago the other two groupings always wanted to marry into the Takudh as they had skills the other two didn't have. These skills were acquired simply because they had the resources to develop them. Examples of these would be tools for hunting, caribou corrals, toboggans, snowshoes, etc.

This story is from the Gwichin, and I offer it in honor of an old friend of mine, Mr. Joe Netro. He shared this story with me during a time in my life when I was having a lot of personal problems and struggles. It was a tough time for me as I was still young and inexperienced. Over a cup of tea while I was visiting with him and his wife, Hannah, he started to tell me this story.

A long time ago, there was an old man and woman. This couple lived alone. Now you might think this unusual in a traditional community, but the reason that they lived alone in their old age was that the woman was barren and could never have children. Well, this winter came really fast. Everything froze solid, the lakes, the river and there was so much snow that even the berries got covered up. And the caribou even took a different trail that year to avoid the storms of their homeland. So, the people had no meat or fish. There was nothing. The land was just blowing snow and ice, and the couple was starving. The woman was in worse shape than her husband. Her eyes were sunken, her belly was swollen, and even the soft spot on the top of her head was sunken in. She was in pretty bad shape. Her husband was starving too, but he was not in such bad shape. He was very worried about her though. Every morning they would say prayers and beg the Creator to give them some food to tide them over until spring. Everyday the old man would go out hunting and return with nothing.

Now on this certain day, the couple had offered their prayers, and the old man realized that if he didn't get anything soon his wife was going to die. Her image was firmly implanted in his mind's eye as he trekked out on the land that day. The wind was howling and blowing over the barren land of rock and ice. The only thing that was out there that day was a small bird, a chickadee, or what most people refer to as a Canadian Jay or camp robber. That's all that was out there, on the land, that day. Now the man wondered if it was the right thing to kill this bird for food, because according to tradition, this bird is the hunter's helper.

When I was a young girl, my grandmother would make me put out chunks of fat out for these particular birds. She would make me do this in the fall just before hunting season. "You hear that guy? He says, "Gimmee fat, gimmee fat!" You have to feed him so he can bring moose to us. He's our helper that guy!"

Now this was the same bird and the man is trying to decide what to do, when he remembers his wife's starving condition. He takes an arrow out of his quiver; the type of arrow which has a blunt end. This is the kind of arrow head which was used for killing birds. The old man takes careful aim and takes the bird down. Carefully he places it into his packsack, all the while giving thanks. Quickly he walks back to his camp to share the good news with his wife.

"The Creator, he took pity on us today, look what I have here!" He showed his wife his catch and she, proceeds to clean and prepare the bird for cooking. She made soup. It has been said that if you are starving, it is not wise to eat solid foods, but to start off slowly, eating soups and drinking tea first; later you can eat a little solid foods. So this is what they did. They drank the soup, and the husband ate a little meat so he could have the strength to go out the next morning. "Let's put this one away for tomorrow," the old woman told her husband. She carefully wrapped the little drumsticks up and put them away in her food cupboard.

The next morning the old woman awoke a little stronger. As they were offering prayers, she was thinking about what a good man her husband was and hoping he was going to get something on his hunt today. Just as he was going out the door, she stopped him.

"Hey, you better take these drumsticks here, you need your strength!"

"No! you better keep it and make more soup, you need it more than me. Look at you!"

They argued about this, but finally the man just went – without the food.

It was only a short distance away that he came up upon two moose, a cow and her two-year old calf. The old man was excited, but calmed himself to take careful aim. He got both of them. He started to clean them, but in his weakened state the only things that he could manage to take home were the kidneys and the hearts. These are delicacies for old people and are usually the first things that are eaten when they take a moose or a caribou.

His wife was so happy! She proceeded to prepare the kidneys and the heart for cooking. It smelled so good! She was looking forward to drinking the broth from this preparation. Her husband was just about to dig in when she said, "Just a minute, before we eat this food, we better eat this one." She pulled out the two little drumsticks she had put away the night before. "That's him, he gave up his life for us. He brought us this moose. We got to give thanks and show respect for him." She placed the two drumsticks before her husband. They ate them slowly, together, and reflected on their blessings.

This is the story of a good woman.

So let's take a look at this story. What are the lessons that we can glean from this legend? First of all let's look at the couple's relationship. They were a team, one balancing the other, interdependent. Relying on each other for support, even at the point of impending death. The husband saw how his wife was suffering more than he was, so he naturally took up the stronger role to take care of her. The fact that they were alone is another point. In most communities of First Nation's elderly people, grandmothers and grandfathers usually have their children take care of them in their old age. Being alone makes the scenario even more vulnerable, and the determination of the old couple becomes even more evident. Their strong spiritual strength and conviction of faith that something good would come as a result of their suffering is a lesson that we can all apply to our everyday lives in so many ways.

So this story is an example of how interdependence went beyond the human to human but also to the interdependence on the spiritual world for assistance. The bird was another level of this interdependence. He was their spirit helper. He was the bringer of game. He also sacrificed himself to ensure their continued existence. Interestingly the old man was put into a position of having to make a decision of whether or not it was wise for him to take the life of this special bird because of its special relationship with bringing moose and caribou. Sometimes we have to make decisions which require sacrifice. When it will have a significant impact on the well being of another individual, then we must make hard decisions. There is so much in this story and a lengthy analysis could be developed, but what is the most important is left to the listeners themselves. It is within yourself. The same holds true for the storyteller as well. Each time I tell and hear the story it means something else for me, too.

2. This story is a solemn one. In that spirit of quiet reflection, invite everyone, in turn around the circle, to share a comment about the story, and how it may have touched them. In the spirit of Louise's own comments, this is not meant to be analytical or argumentative. Rather, this approach respects the reality of different meanings for each listener. To avoid having individuals "feeling on the spot" when their turn comes, participants should be permitted to "pass", meaning that they are not ready to speak and the opportunity to speak will pass to the next person. When everyone has had a chance to speak, those who opted to pass can be invited for a second time to share a thought.

3. Now, have everyone find some bit of material from the site, preferably something non-living and organic, that they could write a small message on. This could take the form of a brief reflection, a short poem, or for those so inclined, a prayer, that conveys something that each participant receives from the earth and is thankful for. When this message of thanks is completed, instruct the participants to find a place to leave it face down. If undisturbed, the message will eventually be composted back into the earth. This is the end of the activity.

## EXTENSION ACTIVITY

It is quite likely that the story above and associated activity will not feel quite right in many locations around the world. That seems to underscore connections between stories, physical places, and activities that occur in them. In fact, the activity above was inspired by another activity in a book called *Keepers of the Earth: Native American Stories and Environmental Activities for Children*, by Michael Caduto and Joseph Bruchac.[9] As written, the activity revolved around a story and a related activity. However, their story seemed out of place in the landscape where our activity above was developed. When an alternative, more place-specific story was selected, the reading of the story seemed to resonate more. However, in addition to reading the story to groups of learners, a short, personal, and reflexive/reflection activity was required. In other words, it was important to listen, but also to physically do something, to give the reflections and feelings a concrete presence in the world. In this case, the original activity inspired a rewritten version that was more attuned to the context.

4. Now it is your turn. How does the activity presented here work in your context? If not, what stories from your landscape would work better? What adaptations would you want to make to the activity? With these questions in mind, use our activity as a springboard to write your own.

## SOME THOUGHTS ABOUT THEORY

### *Locating ethics in a storied world*

One way to locate environmental ethics in the 21st century would be at the intersection of oral traditions, literature, and scholarly inquiry of an emerging field of study. In many ways this is exactly the approach taken throughout this book. To take this framing a step further, we would like to acknowledge those that have gone before us, temporally and culturally. In particular, we have all been influenced by stories from those whose roots are in traditional oral cultures. One such journey follows:

Twenty-five years ago, I moved to the Yukon and began teaching in a small rural school. Most of the students were First Nations. It was a challenging job, but one that provided many lifelong lessons and I am indebted to many people for these opportunities. One important teacher was Mrs. Lucy Wren, a community elder who came to the school to teach the Tlingit language, traditional crafts, and to tell her stories. She was also my first environmental ethics teacher – though she wouldn't call it that.

---

9    Caduto, Michael and Bruchac, Joseph. *Keepers of the Earth.*

Mrs. Wren told many stories but I was most moved by the one about how owls came to be as they are today. She told me how, in the old days, owls were much larger than they are now and how, in difficult times, they could threaten children and old people when other food sources weren't available. This story featured conversations with animals, in this case an owl. It also featured a struggle between a wise old woman and a threatening owl. The result was, through some cleverness and trickery by the old woman, the burning of the big owl. But that wasn't the end. As I recall the story, the ashes from the burning owl ultimately became the small owls of today. There are of course many layers to this story that I don't remember and meanings that I don't understand. And, it is Mrs. Wren's story to tell. But this brief snapshot of recollections provides a starting point for my story.

For me, the work of this story really began when it was contrasted with another story in the schools' curriculum. As it happened, this other story was the tale of St. George slaying a dragon and rescuing a young princess. What Mrs. Wren's story did was to enable me to see a story from my own cultural heritage in a new light. At the time I saw the destruction of a dragon contrasting with a more relational, and accommodating, experience with the owl – the dragon is slain while the owl lives on to learn from the encounter. It struck me that we, at least sometimes, tell different stories, a thought that I learned later was expressed the same way by another Yukon elder, Mrs. Annie Ned.[10] It was my first glimpse at Western cultures' anthropocentrism (or human centeredness), though it would be another year before I discovered this terminology.[11] It was also my first glimpse at storytelling as ethics, though that thought has also taken some years to evolve.

Key to this evolving understanding of storytelling came from another Yukon elder, Mrs. Angela Sidney. According to Julie Cruikshank,[12] Mrs. Sidney was particularly interested in the work that stories do. She was interested in relationships within the stories, but also how the stories help to construct relationships. For Mrs. Sidney we tell stories (at least in part) so people have something to think with. On hearing these reflections, I was again challenged to think about storytelling as research, and as a kind of ethics inquiry.[13]

With the preceding as background, it should be no surprise that stories, storytelling, and the work that stories do, will provide an organising theme for this book. There are two caveats that we must mention, however. First,

---

[10] Carol Geddes, "What is a good way to teach children and young adults to respect the land?"

[11] Another layer of interpretation and reflection can also be explored by considering the religious imagery often ascribed to the St. George and the Dragon Story.

[12] Julie Cruikshank, Unpublished remarks made at the Yukon Teachers' Association Conference, Whitehorse, Yukon, May 1-2, 2003; and Julie Cruikshank, *The Social Life of Stories*.

[13] Bob Jickling, "Ethics research in environmental education," 20-21.

we do not wish to imply that all the relatively recent Western or European stories suggest this sort of contrast with traditional or Indigenous stories. We are speaking here about the work of particular stories. Second, we often find ourselves talking about narrative expression and narrative inquiry at this point in a discussion. However, here we are reminded of David Abram who said, "As far as I can tell, narrative and story mean exactly the same thing. It's just that you kind of need an advanced degree to know what narrative means."[14] We are sympathetic to his perspective, and, since this paper begins with reference to First Nations Elders and their stories, and since theirs is a world of storytelling, we will continue to talk about storytelling. Why not? Storytelling is a perfectly good term, more respectful to those whose insights we have valued and shared, more accessible, and more aesthetically pleasing – to us at least.

## Literature and ethical deliberation

Like oral stories, literature has also informed ethical deliberations. Novels, plays, and poems, for example, add historical and cultural context, challenging conundrums, and thus add nuance to the complexities in which ethics are grounded in the real world. For example, the 19th century Novel by Victor Hugo, *Les Misérables*, tells of the character Jean Valjean initially convicted of stealing a loaf of bread to feed his starving sister and her family. After his release from prison, the novel chronicles Jean Valjean's deeds and misdeeds and the unrelenting pursuit by police inspector Javert. While it is true, in 19th century fashion, that Valjean is cast as the hero in this novel, the reader is taken into a turbulent period of French history replete with moral conundrums that challenge ideas about ethical truths and universal moral imperatives. At a time when categorical imperatives were used to convey a clear and unconditional sense of duty (e.g., stealing is wrong; therefore, stealing food is wrong, even in the face of starvation), Hugo opened a window to the complexities of morality in a real-world context. In a similar way, postcolonial literature has laid bare the political, economic, and social circumstances that have influenced life in the former colonies.[15] Here, writers have mainly used story to provide political and social-ecological critique using the emotive power of their works to instigate political and economic reorganisation of society in the interest of the oppressed. Some of this work also addresses the importance of human-environment relations.

While literature can seem moralistic, it can also do the work of challenging conventional wisdom and revealing complexities in ethical decision-making often overlooked in moral theorising. If Hugo's works

---

[14] Bob Jickling, Pamela Courtenay-Hall, Edmund O'Sullivan, Ann Jarnet and David Abram "What stories shall we tell?" 292.

[15] See for example the review by Fani-Kayode Omoregie, "Rodney, Cabral and Ngugi".

seem too self-assured for the 21$^{st}$ century, contemporary cultural artefacts often portray more ambiguity in the moral landscape. In the television crime drama series, *Breaking Bad*, the audience is constantly being challenged by complex ethical relationships between the law and human circumstances, and the nurturing and demise of the characters' moral impulses. Postcolonial literature, too, has constantly sought to engage the deep-seated ambiguities and ethical contradictions of colonialism, the independence period, and more ethical contradictions embedded in neo-colonialism, especially the ongoing extraction of natural resources from the former colonies. Like stories in the oral tradition, this is the work that literature can do.

While environmental literature in the global North is often associated with the likes of Thoreau and Leopold, contemporary literature with social-ecological themes abound in cultures around the world. A careful reading of works emergent from the global South and from First Nations people show long-standing relational engagements between the human and more-than-human world, as shown in the stories shared above. We encourage readers to use this ever-increasing diversity of titles to help enrich the work of environmental ethics and education.

One way to extend engagement with story and literature as a source for environmental ethics is to join a national or international "Ecocriticism" network. Ecocriticism is a new area of study in the humanities that gained momentum in the 1990s, as more and more literary scholars began to ask what their field has to contribute to our understanding of the unfolding environmental crisis. Ecocriticism involves the study of how cultures – and their stories – construct and are in turn constructed by the more-than-human world. As noted on one website:

> The common ground on which all strands of ecocriticism stand is the assumption that the ideas and structures of desire which govern the interactions between humans and their natural environment (including, perhaps most crucially, the very distinction between the human and the non-human) are of central importance if we are to get a handle on our ecological predicament.[16]

There is also a *Journal of Ecocriticism* which carries interesting articles on how literature and story can bring human, or more-than-human, relations to the fore.[17]

---

[16]  Hannes Bergthaller, "What is Ecocriticism?" *European Association for the study of Literature, Culture, and Environment*, accessed June 19, 2019, https://www.easlce.eu/about-us/what-is-ecocriticism/
[17]  See for example *The Journal of Ecocriticism*, accessed June 19, 2019, http://ojs.unbc.ca/index.php/joe/index

## The emergence of environmental ethics

Traces of what we now call "environmental ethics" can be seen in stories and literature. However, its emergence as a field of study is recent. It has been pursued rigorously as a new branch of philosophy with all the jostling, controversy, and contestation typical in scholarly worlds. In this section, we consider the evolving nature of environmental ethics when viewed as another collection of stories.

Environmental ethics, as a formal field of study, has been around for just over 40 years – that is, as marked by the inauguration of its first journal.[18] Over this period much interesting thinking has accrued and insights gleaned. Interesting theoretical positions have been worked out, challenged, defended, and reworked. However, the field has been criticised too. Some critics, suspicious of meta-narratives and universalising theory, have argued that environmental ethics as a field, is highly moralistic and unreflexive about the full extent of its normalising effects.[19] There is much to consider here.

Interestingly, however, one of the field's predecessors anticipated these very discussions. Aldo Leopold, in his *Sand County Almanac*, is aware that ethics is importantly a process. He spoke of it as an ongoing social evolution, and one that never stops. Implicit in his descriptions are participants who are constantly engaged in the reworking of relationships between themselves and, as he termed it then, the land. Implicit also, is an eschewing of any presumptions of a "true" or "final" ethic. Then he adds,

> I have purposely presented the land ethic as a product of social evolution because nothing so important as an ethic is ever "written" ... It evolves in the minds of a thinking community."[20]

Seen this way, ethical theories are stories that also do work. In this case, they help us in the collective and personal task of working out and living in good relationships with the rest of the world. They are provisional, and useful, but not abstract and universal.

Jim Cheney takes this a little further. He affirms that,

> theories are deeply (though implicitly) shaped by personal and cultural values and that these values are deeply (though again implicitly) shaped by stories – that is, they are carried and propagated by the stories that define us as individuals and define the cultures within which we live and come to understand ourselves.[21]

---

[18]  The word "field" is used in a provisional way. We recognise that the boundaries are permeable and that environmental ethics, thought of this way, rests on preceding histories and oral traditions. See also: Eugene Hargrove, "After twenty-five years".
[19]  See, for example: Éric Darier, *Discourses of the environment*.
[20]  Aldo Leopold. A *Sand County Almanac*, 263.
[21]  Jim Cheney, "The Journey Home," 153.

We agree with Cheney that our understanding of the implicit stories that define us, are not examined often enough. Seeking to reveal some of these underlying stories will be the subject of Chapter 3. Chapter 4 will pick up on the idea that theories can be seen as stories too. Here we will examine some complex questions and quandaries in environmental ethics and some of the influential theories in environmental ethics. We will look at some of their shortcomings, but also some of the work that they do.

## References

Bringhurst, Robert. "The tree of meaning and the work of ecological linguistics." *Canadian Journal of Environmental Education* 7, no. 2(2) (2002): 9-22.

Caduto, Michael and Bruchac, Joseph. *Keepers of the Earth: Native American Stories and Environmental Activities for Children.* Golden, Colorado: Fulcrum Publishing, 1988.

Cheney, Jim. "The Journey Home," in *An Invitation to Environmental Philosophy*, ed. Anthony Weston. New York: Oxford University Press, 1999: 141-167.

Cruikshank, Julie. Unpublished remarks made at the Yukon Teachers' Association Conference, Whitehorse, Yukon, May 1-2, 2003.

Cruikshank, Julie. *The Social Life of Stories: Narrative and knowledge in the Yukon Territory.* Vancouver, B.C.: UBC Press, 1988.

Darier, Éric. *Discourses of the environment.* Oxford: Blackwell, 1999.

European Association for the Study of Literature, Culture and Environment. "What is eco-criticism?" accessed June 19, 2019, http://www.easlce.eu /about-us/what-is-ecocriticism/

Geddes, Carol. "What is a good way to teach children and young adults to respect the land?" in A *colloquium on environment, ethics, and education*, ed. Bob Jickling Whitehorse: Yukon College, 1996: 32-48.

Hargrove, Eugene. "After twenty-five years." *Environmental Ethics* 26, no. 1 (2004): 3-4. https://doi.org/10.5840/enviroethics200426137

Jickling, Bob. "Ethics research in environmental education." *Southern African Journal of Environmental Education* 40 (2005): 20-34. https:// doi.org/10.1080/00958960309603496

Jickling, Bob, Courtenay-Hall, Pamela, O'Sullivan, Edmund, Jarnet, Ann and Abram, David. "What stories shall we tell?" *Canadian Journal of Environmental Education* 7, no. 2 (2002): 282-293.

Journal of Ecocriticism, accessed June 19, 2019, http://ojs.unbc.ca/index .php/joe/index

Leopold, Aldo. A *Sand County Almanac: With essays on conservation from Round River.* New York: Ballantine, 1966. First published by Oxford University Press, 1949.

Maathai, Wangari. *The Green Belt Movement: Sharing the approach and the experience.* Lantern Books, 2003.

Omoregie, Fani-Kayode. "Rodney, Cabral and Ngugi as Guides to African Postcolonial Literature," *Postcolonial web*, 2002, accessed January 2016, www.postcolonialweb.org/africa/omoregie11.html

Profeit-Leblanc, Louise. "Transferring wisdom through storytelling," in *A colloquium on environment, ethics, and education.* Bob Jickling, ed. Whitehorse: Yukon College, 1996, 14-19.

Thoreau, Henry David. *Walden.* New Haven, CT: Yale University Press, 2006.

Shiva, Vandana. *Making peace with the earth.* London: Pluto Press, 2013.

Shiva, Vandana. *Earth democracy: Justice, sustainability and peace.* London: Zed Books, 2006. https://doi.org/10.5040/9781350219755

# CHAPTER 3

## WHAT ARE THE REAL AUTHORITIES IN OUR CULTURES?

**Pedagogical intent:** In this chapter we introduce activities that make the invisible more visible through developing critical literacy and critical analytical skills. In particular, we explore "taken for granted" cultural assumptions using critiques of everyday artefacts. This chapter will also illustrate how important things in our lives can be devalued and reduced in value through contextualised pressures that are seldom recognised.

**Activities:**    3.1: Investigating unquestioned assumptions

3.2: A curriculum critique

3.3: Analysing self-validating reductions

3.4: Reducing the world

3.5: Stories of reduction

3.6: Locating space for resistance

**Main theoretical resources:** Elliot Eisner's explicit, implicit and null curriculum theory, Anthony Weston's self-validating reduction and his self-validating invitation inform this chapter. It also draws on work by First Nations Elders, Neil Evernden, and David Abram.

## Being critical

### Rare bugs, advertising, and unquestioned assumptions

In August 2002, talk about the World Summit on Sustainable Development (WSSD) in Johannesburg was everywhere. During the early days of this conference, the pages of major Canadian newspapers presented an array of perspectives. Some authors were more positive than others, but there was an overall sense of anticipation, and even hope. Big issues like global climate change and biodiversity were discussed.

Many Canadians had been thinking about biodiversity and they read, with interest, prospects for pressing this agenda forward. At one point in time, they had been lobbying hard for endangered species legislation. Then, some opinion polls indicated that as many as 94% of Canadians had urged the Minister of the Environment to give their country strong legislation in this area. Many found comfort in these numbers. They saw a strong undercurrent of environmental concern and, in particular, concern for disappearing species and habitats that support them. Reading the newspapers conveyed a renewed sense of optimism. It was good to know that biodiversity was being included in mainstream media reporting as the second Earth Summit (2002) was getting underway. However, such optimism can be fleeting. Consider the Air Canada in-flight magazine *En Route*[1] that was published at the same time.

Prominent amongst other marketing in this magazine was an advertisement for a four-wheel drive vehicle sitting in the middle of a jungle stream. It looked something like this:

Figure 3.1    Subaru. "Start a Rare Bug Collection on Your Windshield." Advertisement. En Route Magazine.[2]

---

[1]    See Air Canada, *EnRoute Magazine*.
[2]    Ibid.

The caption presented below this advertisement read: "Start a rare bug collection on your windshield." Further below, in the small print, it said, "Entomologists aren't the only ones who can discover a new insect. All you need is the ..."

Here, marketers invoked the collectability of rare insects – or even worse, the destruction of rare insects by their products' windshields – as a sales gimmick. What's more, their pitch is made to a community that supports strong endangered species legislation. Marketers are clever; they know how to sell products. Yet, this advertisement flies in the face of public sentiment. Why?[3]

Canadians are not alone in such apparent confusion. Here are two other examples that seem to illustrate the same paradox.

### A natural high?

A Scottish tourism flyer encouraging visitors to experience a "natural high" in CairnGorm mountains tourism project tells a similar story. The flyer advertises the "inspiration, adventure, and enjoyment" of mountains as an enticement to patronise the new mountain railway that now carries visitors "in total comfort (protected from the infamous Scottish weather!) from car park level to the new Ptarmigan Station and the higher lying snowfields in around fifteen minutes." On arrival, this Ptarmigan Station offers "spectacular views combined with a mountain exhibition and the highest panoramic restaurant and shopping in the UK."

There are many ways to analyse this Scottish flyer. One can, for example, immediately see the juxtaposition of contrasting, even contradictory, messages. How does one reconcile the inspiration of the mountains – the natural high – with the total comfort and protection of the artificial environment of the railway car? Then, there are unforgettable mountain memories that are topped off with dining and shopping.[4]

### Possessing rarity?

Another interesting story is to be found in a special edition of the South African Airways in-flight magazine, Sawubona, published during the World Summit on Sustainable Development (alternatively, "The Second Earth Summit") to draw attention to sustainable development. It also carried an advertisement for an extremely rare stone, the Tanzanite.[5] The headline read "At last it is acceptable to wear something on the endangered list." A featured selling point of this rare stone is a trademark that reads "Mark of Rarity." Text in a more recent advertisement for this stone reads: "With only a single known source at the foothills of Mount Kilimanjaro, and just

---

[3]   This story is taken from Bob Jickling, "Making Ethics an Everyday Activity".
[4]   Ibid.
[5]   Sawubona Magazine. "World Summit Special," August 2002.

over a decade's supply left, tanzanite is the world's rarest and most mystical gemstone. Insist that your tanzanite comes with the Mark of Rarity – it's your assurance of authenticity, accurate grading and an ethical route to the market."

It seems ironic that a "Mark of Rarity" (or scarcity in other words) is seen as an "ethical route to the market". The hidden message is about private ownership of a scarce resource, or "get your tanzanite before it runs out, and make sure you pay a high enough price for its rarity!"

When extracted from their advertising contexts these examples leave many of us shaking our heads. Why do such seemingly contradictory messages so often go unnoticed? What can be done to make plain the tension between environmental ethics and consumerism, as well as other aspects of popular culture?

## LEARNING ACTIVITY 3.1: INVESTIGATING UNQUESTIONED ASSUMPTIONS

As we can see from the examples above, beliefs or assumptions are often embedded so deeply within the text and images of cultural artefacts, such as advertisements, that they become almost invisible. Hidden this way, they remain unquestioned. These unquestioned assumptions are what environmental philosopher Neil Evernden[6] has called the *real authorities* of a culture. They are also at the root of an ethics problem. It is these assumptions, whether we are aware of them or not, that shape our practices and the decisions we make.

With this in mind, we begin our inquiries by searching for these embedded assumptions – these "cultural authorities". Once revealed, we, and our students, can be more critical as we evaluate priorities and preferred futures. This process is the reflective practice of ethics.

1. Consider the advertisement for the four-wheel drive vehicle sitting in the jungle stream with squashed insects on its windshield. What hidden messages does it contain? How is nature interpreted in these examples? How is nature valued? How are people to relate to nature? Or, to the more-than-human world? Also look closely at the language used. Is it respectful of people, animals, plants, ecosystems, or does it imply that they are simply objects of consumption? What are the consequences of language choices?

---

[6]   Neil Evernden, *The Natural Alien: Humankind and Environment.*

## EXTENSION ACTIVITY

2. Consider the stories "A Natural High?" and "Possessing Rarity". What hidden messages do they contain? What do they reveal about how people are to relate to each other and the natural world?
3. These kinds of stories are easy to find. Newspapers, and other media, can be good sources, particularly if an issue is followed over a period of time. Find some. By collecting a series of articles over a period of weeks, or even months, a range of perspectives can be gathered. What messages do these various perspectives contain? What assumptions do they rest on? With assumptions revealed, discuss preferred options.

## Thinking critically about curricula

Ernst Friedrich Schumacher, in his widely read book *Small is Beautiful*,[7] raised doubts about the efficacy of Western education. Despite widespread belief in education as the key to a resolution of our problems, and despite vast amounts of energy and resources devoted to education, Schumacher suggested that little had changed.

Forty years after Schumacher's observations, there were some indications that environmental education may be having some impact. Perhaps most telling is that the corporatist community is lashing out against the field. For example, the *New York Times* reported: "The environmental education movement is based on flawed information, biased presentations and misguided objectives. At worst ... impressionable children are being browbeaten into an irrational rejection of consumption, economic growth and free market capitalism."[8]

While attacks on environmental education are increasing, corporatists have also sought to exert their influence through development of school curricula and teaching aids. These often support consumerism and models for economic growth that Schumacher warned us about.

An interesting example emerged in the Yukon Territory in Canada after a group of school children published letters in a local newspaper. The letters expressed concerns about the future of mining in the Territory and suggested rethinking approaches to this industry. The mining industry reacted negatively and attacked the teacher, the school, and the curriculum as being biased. They concluded that remedial action was required and that, together with the Department of Education, they should develop a "Mining Curriculum" for Yukon schools. The bluster surrounding this issue was played out through several letters and articles published in the local papers.

---

[7] Ernst Friedrich Schumacher, *Small Is Beautiful.*
[8] John Cushman, "Critics Rise Up Against Environmental Education."

One of the most interesting articles was an editorial entitled "Corporate money is seductive, but do we want it?"[9] Here, the editor acknowledges that those preparing study materials would be professionals and would not easily be bamboozled into planting uncritical and subliminal industry messages into the curriculum, but he still speaks of nagging doubts about this kind of funding arrangement. He concludes that corporate sponsorship in the public school system is a seductive and dangerous practice. This editorial leads to important questions for anyone interested in curriculum developments.

How, for example, can educators see through attempts to bamboozle them? Some supplemental curricula and teaching materials are very sophisticated. How do curriculum sponsors subtly, and perhaps even unwittingly, influence projects they pay for? What assumptions about contemporary society, economics, and human/nature relationships do these curriculum projects rest upon?

Hidden assumptions are not only found in advertising examples, but in all curricula. What assumptions do these curricula represent? Do they encourage us to explore viable alternatives for the future, or do they tend to reinforce the status quo? There will always be values embedded in any teaching materials – some openly stated, others implied through the text, and yet others revealed through strategic omissions.

Educators have recognised for some time that no curricula can be value-free. Among them, curriculum theorist Elliot Eisner[10] has provided a framework that can be used to critique educational materials. In a chapter called "The three curricula that all schools teach", he argues that there will always be an explicit, an implicit, and a null curriculum. In revealing these three curricula we are also revealing the beliefs and assumptions which shape learning experiences and educational materials. Eisner's framework, used as a critical tool, can provide a basis for accepting or rejecting curricula, or for using a curriculum in a thoughtful and analytical way – as a vehicle for revealing and examining social assumptions, as a basis for making more informed judgments, and as a factor in justifying our decisions.

---

9    Ken Bolton, "Corporate Money Is Seductive, But Do We Want It?"
10   Elliot Eisner, *The Educational Imagination.*

# HANDOUT: THE THREE CURRICULA

*Some more information on the "Three Curricula"*

**The Explicit Curriculum** – This refers to the curriculum defined by the stated goals and objectives.

**The Implicit Curriculum** – This refers to the "hidden curriculum" or the curriculum that is not openly stated yet is required by virtue of the content selected or the teaching practices used. Thus, the social roles and relationships in classrooms, and the content selected, carry "hidden messages". A detailed, highly structured programme with tight time allotments implies that both teachers and students need a lot of direction and control, and that learning is best accomplished in a regulated environment. Or a classroom that respects children and encourages them to voice their opinions carries a hidden message that children are valued. The scope for decision-making, for student-teacher interactions, for criticism, may all be elements of the "implicit curriculum". The implicit curriculum is also represented by the values and assumptions embedded in the language, images, and examples employed.

**The Null Curriculum** – This curriculum is defined by what is not said, discussed, or included. Often it reflects basic political decisions made during the process of curriculum development. Suppose, for example, that a major forestry corporation was to support a curriculum in forest management. If the final programme made no mention of Indigenous people's land claims, cultural practices, and rights, then that set of ideas would be part of the null curriculum. Most likely they would have been omitted because sponsors decided that they were not acceptable or that they were irrelevant. What is *not* said often says more about a curriculum than what is said. It is interesting to see if particular perspectives are promoted, either overtly, or through a failure to include and examine alternative perspectives.

## LEARNING ACTIVITY 3.2: A CURRICULUM CRITIQUE

One outcome of the Yukon mining story was an alliance formed between industry and government to produce a mining curriculum called *Rock on Yukon*.[11] One element of this project was a poster to advertise its arrival into schools and to capture the interest of the targeted children. It looked something like this:

Figure 3.2    What's mined is yours (Photos by author, Bob Jickling)[12]

The first thing an observer notices is that the image is full of happy people enjoying products derived from mining activities. These images compliment a long list of useful products that are derived from mining activities. A thematic slogan runs across the poster and reads, "What's mined is yours". Of particular interest is an additional text that states, "At the heart of our modern lifestyle is a diverse and healthy mining industry."

1. Using Eisner's three curricula and the accompanying handout as a guide, take a close look at this poster. What kinds of images are used to support the curriculum? What hidden messages does it contain – that is, what are the implicit messages? What do they reveal about how people are expected – or conditioned – to relate to each other? How are people expected to relate to nature? Or, to the more-than-human world? And importantly, what isn't there? What assumptions are hidden by omission – by virtue of the null curriculum? Are there assumptions that you expect the designers did not want to have revealed or challenged?

---

[11]    Jeannie Burke and Eric Walker, *Rock On Yukon*.
[12]    These images were taken of a well-used lesson aid that accompanied the text by Jeannie Burke and Eric Walker, *Rock On Yukon*.

2. Activity 3.1 in this section can help to sharpen up your analytical skills and point to inevitable value-laden positions in any curriculum. However, the sample investigated arises in response to public controversy and an intervention into that controversy by those with vested interests. You might ask, does the same thing happen in government-supported curricula? What do the "authorised" versions of curricula look like?

3. 3. In 2011, the South African Government undertook a curriculum renewal project and released new curriculum and assessment statements. The document includes specific aims that purport to define the purpose of studying life sciences. A few pages along, some "investigations" are provided that are intended as ways to apply life science learning in contemporary contexts (See the handout titled, "The Specific Aims of Life Sciences").

Again, use Eisner's three curricula to guide your investigations. The three statements about the purposes of studying life sciences constitute the explicit curriculum. As you look more closely inside of these explicit statements, what values are implicit, or implied? What topics are left out, or null?

Some things that interested us included the emphasis on objective thinking? What do you think of this? What does this imply about the kinds of knowledge that are important in the school system? What might be left out? What aspects of life sciences are not problematized?

Another interesting statement in the handout links science, decision-making, and ethics: "Moreover, understanding science also helps us to understand the consequences of decisions that involve ethical issues." On the one hand, connections such as this are uncommon in school science curricula. Ethics are often part of the null curricula. However, if you push a little further, what does your analysis of implicit and null curricula reveal? Objective thinking is emphasised in the second aim above. Does the third aim, stressing relations between science and human ability, open spaces for more subjective understanding of the world? Now look at the suggested investigations that follow the aims. What do you like about them? What is implied? Can you identify null aspects of these investigations? Do they support the aim to further understanding of more subjective elements in relations between science and human ability? Or not?

Reflecting on, and discussing, the questions in this activity can be productive and illuminating. Good luck. We will come back to topics raised by these questions again in Chapters 6, 7, and 8.

# HANDOUT: THE SPECIFIC AIMS OF LIFE SCIENCES

*(Below is an extract from the South African National Curriculum Statement indicating the three specific aims of the subject of Life Sciences)*[13]

*The development of Scientific Knowledge and Understanding*

Scientific knowledge and understanding can be used to answer questions about the nature of the living world around us. It can prepare learners for economic activity and self-expression and it lays the basis of further studies in science and prepares learners for active participation in a democratic society that values human rights and promotes acting responsibly towards the environment.

*The Development of Science Process Skills (Scientific Investigations)*

The teaching and learning of science involves the development of a range of process skills that may be used in everyday life, in the community and in the workplace. Learners can gain these skills in an environment that supports creativity, responsibility and growing confidence. Learners develop the ability to think objectively and use different types of reasoning while they use process skills to investigate, reflect, synthesise and communicate.

*The Development of an Understanding of Science's Roles in society*

Both science and technology have made a major impact, both positive and negative, on our world. A careful selection of scientific content and the use of a variety of methods to teach and learn science should promote the understanding of science as a human activity as well as the history of science and the relationship between Life Sciences and other subjects. It also helps learners to understand the contribution of science to social justice and societal development as well as the need for using scientific knowledge responsibly in the interest of ourselves, society, and the environment. Moreover, understanding science also helps us to understand the consequences of decisions that involve ethical issues.

*(Below is an investigation proposed in the 2011 South African Life Sciences curriculum)*

Case study: Rationale for culling, e.g., elephants in the Kruger National Park as an example of an application of estimating population size (link to researched reasons for culling).

Draw up a public survey form to test the public opinion about culling. Show results in a pie graph.[14]

---

[13] Department of Basic Education. *Curriculum and assessment policy statement*, 12.
[14] Ibid., 49.

## EXTENSION ACTIVITY

It is not possible to judge an entire curriculum by critiquing just a poster or its aims. It is a start, but to be fair you need to examine the whole package. So, try this:

4. Find a curriculum, or some curriculum materials that you are interested in – perhaps something from your region – and take a close look. What is the explicit curriculum? What does it say it will do?

    What is implicit? In addition to looking at the stated goals, check out the activities. What hidden messages do they contain? What do they reveal about how people are to relate to each other and how they are to relate to nature, or, to the more-than-human world? Look closely at the language used. Is it respectful of people, animals, plants, ecosystems, or does it imply that they are simply objects for consumption? What are the consequences of language choices? What kinds of images are used to support the curriculum? What assumptions are embedded in their selection? Is nature portrayed as a resource to be used, a gymnasium, a playground? Is nature portrayed with only picturesque images? Or images without people? How are people's cultural values portrayed? How are people-environment relationships portrayed?

    What isn't there? What is the null curriculum? What seems to be omitted for political or other reasons? Moreover, what does this tell you about the curriculum?

    These questions lead to judgments that educators must make in deciding to use curricular materials. You might ask yourself: Is the curriculum generally educative, or does it advocate a particular view? Does it attempt to initiate the student into a particular set of social norms? Does it encourage student understanding, critical thinking, and judgment, or does it attempt to train the student, or modify the student's behaviour in a particular predetermined fashion?

5. In Chapter 1, we discussed the rise in prominence of UNESCO's 17 Sustainable Development goals and their associated targets.[15] We suggested that, given their profile that they warrant both consideration and scrutiny. The analytical tools discussed in this chapter can be used to take a closer look at these goals and their targets. What is explicit, implicit, null? How is nature valued? What kind of human/nature relationship is implied?

## Self-validating reduction

The activities and stories in this section reveal what can happen when the potential of people, communities, places, or the more-than-human world is reduced in a downward cycle, or in a way that is "self-validating".

---

15  See for example, UNESCO, "Sustainable Development Goals."

This process has also been called "self-validating reduction", a concept coined by Anthony Weston. It is a challenging concept, but this section describes, in more detail, how it works and gives examples that show how "self-validating reduction" can be used to probe deeply held cultural assumptions and to open everyday ethics questions.[16]

## What is self-validating reduction?

A good way to begin is by exploring the consequences of reducing potential. In this way, self validating reduction is like a self-fulfilling prophesy. The more you expect something to happen the more it will happen. The classic example is treating underperforming students as if you think they are dull or stupid. If teachers send them that message – maybe by humiliating them in class or telling them that the work is too hard for *them* – these students begin to take on "traits of being stupid". The self-fulfilling part kicks in now. As students begin to believe they are not very smart, they underachieve even further. This, of course, "validates" the teachers' assessment – in their own minds – and seems to justify yet more demeaning comments. In Weston's words, "the circle closes".[17] In this scenario the student can be doomed.

Self-validating reduction describes a similar reducing cycle. In these cases, however, it is typically something physical and real in the world that is reduced. Take chickens for example. In their free-roaming state they are beautiful birds that can give people pause. They can appear dignified – maybe regal – and worthy of respect and care. However, when they are bred specifically for meat, they can become disproportionately heavy, and a lot less attractive. This less attractive bird may, in turn, invoke less respect, thus, making it easier to continue to breed even higher meat-yielding birds. The resulting birds can be so badly proportioned that they are unable to support their own weight. As such, they become pitiable and grotesque. With less respect garnered, it becomes ever-more easy to raise these "disgusting" birds in factory-like conditions. And so it goes.

The same kind of thing can happen to physical spaces. If there is a big piece of "undeveloped" land, in some places it might be called wilderness, elsewhere it might be traditional territories for Indigenous peoples, or maybe wildlife preserves. Yet other places might be considered to be spiritual or holy sites. While these pieces are whole, they have enormous value for a lot of people because of the integrity of this wholeness. However, if a road is built through the middle of such an area, the wholeness is fragmented and diminished. When this once special piece of

---

[16]   This material is based on a workshop activity originally prepared by Anthony Weston. See also, Anthony Weston, *Back to Earth: Tomorrow's Environmentalism*; and Anthony Weston, "Self-validating Reduction: Toward a Theory of the Devaluation of Nature," 115–132.

[17]   Weston, *Back to Earth*, 96.

landscape is diminished, it no longer holds quite the same feel that it once had. Thus diminished, the remaining land is more vulnerable to further development – maybe developing some additional side roads to particular resources. This incremental development is typically framed as small and minimally disruptive. In a way this might be true. Still, the net effect is that this incremental growth of roads makes it more likely that further roads are put in place. So, building the first road, in a way validates more roads, as well as other developments that are needed to support roads such as fuelling stations, stores, etc. More roads validate the building of yet more roads, and on it goes. The place is doomed once the spell is broken and encroachment begins.

*Prejudiced* views readily become self-fulfilling too. Take, for example, a sexist attitude held by a man who views women as "less than" himself. Armed with such an attitude, the man interacts with the women in his life in ways that are degrading, or even worse, abusive. After a while these women may become conditioned to see themselves as inferior, less intelligent, less able or less powerful. Seeing themselves as inferior, they may begin to act that way, confirming the original sexist attitudes. To the man, this behaviour encourages evermore sexist actions, and the cycle becomes self-fulfilling, or even more accurately self-validating.

The same kind of self-validating views are often the basis of racism. One sometimes hears the statement that "racism breeds racism", or consider slavery. A system that deadens people with work and fear does not merely reflect and respond to prejudice: it perpetuates it and begins to *justify* it. Then, the very crimes of slavery – that it "makes the enslaved a character fit only for slavery"[18] – become slavery's best defence.

In cases such as sexism and slavery we need to speak of something stronger than mere "prophecy". Beating or bullying someone into submission, subservience, or slavery, for instance, is no mere "expectation". Here the person is *reduced* to something less: a thing or object, something to be exploited, dominated, or killed. This is where the terms "self-validating" and "reduction" come together. The "reduction" justifies itself to such an extent that it validates itself. We can now speak of self-validating reduction. Exploring examples of self-validating reduction related both to the more-than-human world and to issues of human social justice (e.g., sexism, or racism) is important for environmental ethics, as ethical tensions related to the natural world are rarely free of social justice concerns. Indeed, people worldwide who experience social oppression through racism and/ or poverty bear a disproportionate brunt of the negative results of the ecological crisis.[19] An example of such complex intersecting reductions, and subsequent injustice is illustrated in the example "Small farmers sink or swim" in the participant handout on self-validating reduction.

---

[18]  Frederick Douglass in Weston, *Back to Earth*, 97.
[19]  Kari Norgaard, "Climate denial and the construction of innocence," 80-103.

A self-validating reduction, then, is a self-fulfilling prophecy in which one of the main effects of the "prophecy" is to *reduce* someone or something in the world. It acts to make that person or thing less than they are or could be; or it diminishes some part of the world's richness, depth, and promise. This reduction in turn feeds back, not only to justify the original "prophecy" but also to perpetuate it. Here are a few environmental examples of self-validating reduction:

The treatment of animals in factory farms where their social instincts are destroyed – this is where chickens are debeaked and pigs have their tails docked – and they have no chance to develop any kind of intelligence or relationships with each other or human beings. So *of course* they will seem stupid, slothful, pitiful, or vicious, and deserving of their fate.

Wild rivers are destroyed by dams, rich ecosystems are simplified, clear cut, or strip mined, and so on; then we begin to lose a sense of why nature or wilderness has any real value in the first place. This, then, makes it easier to justify further developments. The more development that takes place, the easier it is to argue for ever-more development.

Arundhati Roy[20] describes experts at the World Water Forum at The Hague in Holland discussing privatisation of the world's water. Delegates at this forum believed that "God gave us the rivers, but he didn't put in the necessary delivery systems. That's why we need private enterprise" – privatisation is necessary because there are no "natural" delivery mechanisms. As experts convince themselves of the need for privatisation to provide water delivery mechanisms, others start to believe that this is the only way to create access to water resources. This validates further privatisation of water. Through this self-validating process, water is reduced from a basic need, a free commons resource, and a right, to a commodity that can only be bought. It becomes easy to forget that, in the past, people accessed water in many different ways, freely as a commons resource that belonged to all equally – throughout humanity's long history – and that there may be many other creative solutions, besides privatisation, to the distribution of water resources in modern societies.

One such a solution was found in post-Apartheid South Africa. During the Apartheid era, farmers "owned" all of the water that ran through their properties, which placed most of the water resources in the hands of the rich white farmers. In post-Apartheid South Africa, water was declared a common resource, owned and protected by the state, to be allocated for use via a licensing system; distributional arrangements were worked out from this starting point. The legislation also allocated a percentage of the water to the environment to maintain ecosystems, biodiversity, and the ecological infrastructure.

Another example comes from economic policy. Free trade is "marketed" by dominant economic groups and institutions as the solution to poverty

---

20  Arundhati Roy, *The Algebra of Infinite Justice.*

and development problems. As more political support is given to this policy at national and international levels, it becomes integrated into government policies. With increased support for this economic policy, the policy assumptions become self-validating. Now, many people have come to believe that free trade is the only economic framework that can guide world development – alternative economies are reduced.

## LEARNING ACTIVITY 3.3: ANALYSING SELF-VALIDATING REDUCTIONS

1. In small groups, consider one of the examples given in the handout on self-validating reduction. What reductions are evident in these situations? What possibilities exist for further discounting based on the reductions that you have identified? Teachers may use these examples as a way of introducing the concept of self-validating reduction.

## EXTENSION ACTIVITY

2. To illustrate the logic of self-fulfilling prophecy from the "inside", ask how often we find ourselves unable to escape being reduced to some stereotype. A simple example is clumsiness. If you know someone thinks you are clumsy, you are likely to soon produce real clumsiness. To be graceful, as opposed to clumsy, is partly about being unself-conscious – a state difficult to achieve when you are stereotyped. So, you become more self-conscious, and this leads to more clumsiness, and the cycle repeats itself. What examples of self-validating reduction have you experienced?

3. Turning to reductions of the land and of nature as a whole, identify some of the prevailing images of nature, for example in newspapers, and then ask how they came to be reduced in that way. For example, when land is divided into pieces for sale – readily observed in real estate advertisements – how long is it before the ecosystem itself is divided into pieces, fragmented? When sacred places are turned into holiday resorts, do they retain their power, or do they precisely become holiday resorts? Find examples of these kinds of reductions in newspapers, magazines, and in television and Internet advertisements.

4. Ask participants to brainstorm possible instances of self-validating reduction that are local to their communities, or relevant to their lives. Have students write brief profiles to describe the self-validating reduction that they have identified. Alternatively, they could draw or use other means of artistic expression to report their examples.

# HANDOUT: SELF-VALIDATING REDUCTION

### Small farmers sink or swim

Mexico has a long history of producing and marketing corn. As a result of the North American Free Trade Agreement (NAFTA), Mexico has been flooded with imported corn from north of the border. This has led to a drop in the price farmers received for their corn, and consumers have ended up paying more. The price of tortillas – the country's staple food – rose fivefold as the Government jettisoned domestic subsidies and giant agribusiness firms took over the market. The Government of Mexico recognises that reality did not turn out the way they planned it. A sink-or-swim policy for small farmers led to a drop in income that hit the most vulnerable members of rural society. The remaining Mexican corn farmers now face the additional risks associated with the contamination of native varieties of corn with genetically modified imported maize, which could have major consequences for Mexican *Campesino Farmworker Centers* and for local biodiversity in Mexico. 'Corn is important because it allows us to live in peace … it is our form of food security' says one of the villagers, showing that free trade assumptions affect more than just the economies of developing countries.[21]

### Tranquilize monkeys in the name of science?

Visiting a rainforest research station at which howler monkeys are being studied, Jim Nollman is assured by researchers that the monkeys are (in their view) fundamentally unsociable, retreating to the forest canopy whenever humans are around. This is demonstrably what they have done for years. Then Nollman learns that the zoologists study the monkeys by attaching radio transmitters to their necks. To attach the transmitters they have to tranquilize the monkeys. To tranquilize the monkeys they shoot them with tranquilizer guns, dropping them out of the trees. The zoologists consider this technique an unproblematic objective, purely scientific, and they treat Nollman, a musician who tries to use music to create a shared space between humans and animals, as just a sentimental and unscientific meddler.[22]

Could even the attempt at scientific objectivity produce self-validating reduction?

---

21 Story adapted from Laura Carlsen, "The People of the corn," 12-13.
22 Anthony Weston, *Back to Earth.*

## A bit more about self-validation

Sometimes the reduction aspect of an action can easily be seen, but the self-validation potential can be a little trickier. It can be helpful to think back to the activities in the "being critical" section of this chapter. Think about how an insect splattered on a car window has been used to sell cars. The image of nature presented in this advertisement could be described in many ways such something collectable, a commodity, or a trophy. For you as a reader, this might be a different or new view of insects. This might be a new view for many others, too.

Sometimes this so-called "new view" of something – once it has been reduced – can become so unconsciously embedded in a cultural context that it changes the way that things are seen and done, particularly when the "new view" becomes part of our day-to-day lives and our taken for granted "way of thinking". Sometimes the "new view" – that is the reduced view – is then put forward as "the only view" or "the only way" that things can be seen.

What is important, however, is to be alert to what happens when this often unconscious process reduces the full potential of human beings, or the more-than-human world in such a way that potential can no longer be seen, or that it is changed forever or lost.

With this background, consider the implications of auctioning hunting permits for exotic animals. What could be the consequences of reducing exotic animals to commodities for human consumption? The accompanying handout provides examples of two such events. The first describes the auctioning of a permit to hunt a black rhino, and the second was for a permit to hunt a lion. As it turns out, the lion turned out to be named "Cecil". In the first of these two articles a critic of the rhino hunt said, "This auction is telling the world that an American will pay anything to kill their species." If this is true, what does this say about how a new view about exotic species is beginning to take hold in some parts of the world?

If there is indeed a new view of exotic species taking hold, then an interesting question becomes, can this be self-validating? In other words, do auctions for hunting permits become normalised, making them ever more acceptable? The learning activities that follow will allow you to consider these questions.

## LEARNING ACTIVITY 3.4: REDUCING THE WORLD

1. Consider the two news stories in the accompanying handout; one is about black rhino hunting and the other is about the killing of Cecil the lion. First, what hidden messages are contained in the two stories? How are animals portrayed? What metaphors might represent these

hidden messages? For example, what are animals to the hunter who was described as being "in pursuit of a new bow hunting record?"

2. In answering question 1 above, you may well have identified ways in which the animals involved have been reduced. Now the tricky question to think about is: In some contexts, have these reductions been so widely incorporated into social norms that they hold considerable weight as a "new view?" If so, does this "new view" make it easier to treat ever more animals in similar ways? If you think the answer to the last question is "yes," then the reduction of the animals (to commodities, trophies, or ...) is being self-validated. See if you can come up with other ways in which the self-validating process works in these examples.

## EXTENSION ACTIVITY

3. The idea of legitimising "new views" in a social context, and the potential self-validating effects of these new views brings an important twist to the idea of self-validating reduction. For example, in an era where climate change is on the cusp of becoming a critically important socio-ecological issue, the concept of "junk science" has entered the global discourse. This term is often used by people who are not scientists, and who are typically not in a position to judge what good science would entail. However, the incorporation of the term "junk science" into everyday language suggests that the value and integrity of science itself has been reduced. And, if this reduction becomes part of an emerging new view, then this reduction could be self-validating.

Now see if you can find examples in the popular press of language that reduces the value of science? In relation to climate change? In relation to other environmental topics? In relation to the COVID-19 pandemic? Can you find other examples of how another part of the more-than-human world is similarly reduced by an emerging new view that is subtly being incorporated into everyday discourse?

# HANDOUT: BLACK RHINO HUNTING

*Self-Validating Reduction*

**Black Rhino Hunting Permit Auctioned For $350,000**

January 23, 2014. DALLAS (AP) —A permit to hunt an endangered African black rhino sold for $350,000 at a Dallas auction held to raise money for conservation efforts but criticized by wildlife advocates.

Steve Wagner, a spokesman for the Dallas Safari Club, which sponsored the closed-door event Saturday night, confirmed the sale of the permit for a hunt in the African nation of Namibia. He declined to name the buyer.

The Safari Club's executive director, Ben Carter, has defended the auction, saying all money raised will go toward protecting the species. He also said the rhino that the winner will be allowed to hunt is old, male and nonbreeding—and that the animal was likely to be targeted for removal anyway because it was becoming aggressive and threatening other wildlife.

But the auction drew howls from critics, including wildlife and animal rights groups, and the FBI said it was investigating death threats against members of the club.

Officials from the Humane Society and the International Fund for Animal Welfare have said that while culling can be appropriate in abundant animal populations, all black rhinos should be protected, given their endangered status.

An estimated 4,000 black rhinos remain in the wild, down from 70,000 in the 1960s. Nearly 1,800 are in Namibia, according to the Safari Club.

Critics have also said any hunting of a rhino sends a bad message to the public.

"This auction is telling the world that an American will pay anything to kill their species," Jeffrey Flocken, North American regional director of the Massachusetts-based IFAW, said this past week. "This is, in fact, making a spectacle of killing an endangered species."

The auction took place in downtown Dallas under tight security. Organizers hoped to at least break the previous high bid for one of the permits in Namibia, which was $223,000, and had said the amount could be as high as $1 million. The nation offers five permits each year, and the one auctioned Saturday was the first to be made available for purchase outside of Namibia.[23]

---

[23] The Guardian, "Black rhino hunt permit brings $350,000 at controversial auction."

# HANDOUT: AMERICAN HUNTER KILLED CECIL

### American Hunter Killed Cecil, Beloved Lion Who Was Lured Out of His Sanctuary

Cecil, a 13-year-old lion, wandered out of his sanctuary in a national park in Zimbabwe this month, following the scent of a potential snack. At the other end of Cecil's search was a lure, placed there by hunters who, conservationists say, wanted their prey to cross into unprotected territory so they could kill him. Cecil, well known to those who visited the Hwange National Park in western Zimbabwe for his jet-black mane, was beheaded, according to conservation officials. His corpse was left to rot in the sun.

Zimbabwean officials said that Dr. Walter J. Palmer, an American hunter known for killing big game with a bow and arrow, killed Cecil, and was being sought on poaching charges. Johnny Rodrigues of the Zimbabwe Conservation Task Force said, "Cecil was lured out of a protected game preserve one night in early July by a hunting party that tied a dead animal to a car."

The first shot, which the authorities say came from Dr. Palmer's crossbow, was not enough to kill the lion. Cecil was tracked for nearly two days before Dr. Palmer killed him with a gun. The details of the lion's death have outraged nature enthusiasts and conservationists around the world who are troubled by wealthy big-game hunters who pay tens of thousands of dollars for licenses to kill protected animals for trophies and sport.

Hunting advocates and some conservationists argue that, if done responsibly, the selling of expensive licenses to big-game hunters can help pay for efforts to protect endangered species. In 2013, the Dallas Safari Club in Texas fought for the right to sell at auction a permit for the hunting of a black rhino in Namibia, setting off a debate over the practice. The group argued that a limited hunt helped thin the herd of weak rhinos so the population could grow, and that the $350,000 paid in 2014 by a reality show host to hunt the animal would help fund Namibia's conservation efforts.

In 2009, Dr. Palmer, a dentist from Minnesota, paid $45,000 at an auction to help preserve an elk habitat in California. A big-game hunter who prides himself on his skills in hunting without firearms, Dr. Palmer was profiled in 2009 in The New York Times, when he shot an elk dead from 75 yards with a compound bow in pursuit of a new bowhunting record. The Telegraph in Britain reported on Tuesday that he paid around $54,000 for the opportunity to hunt a lion…[24]

---

[24] Katie Rogers, "American Hunter Killed Cecil."

*Using self-validating reduction to tell a story*

In this chapter, self-validating reduction has been presented as an analytical tool. In this activity, we will explore how we can use this concept to tell a story, suggest alternatives, or issue a warning. Or, to put it another way, how can knowledge of self-validating reduction be used to open perspectives in ethical decision-making?

One thing to consider is that the general public is not familiar with the term. One tactic would be to find ways of making the point using more user-friendly language and metaphors. Consider the article "Privileging prospecting, staking, and mining is a breach of trust!" It was written for a community newspaper amid sometimes-heated discussions over contested mining legislation. The article is edited a little for brevity, but a link to the original article is provided.

## LEARNING ACTIVITY 3.5: STORIES OF REDUCTION

1. After reading the article described on the next handout page, consider how the idea of self-validating reduction is used to tell a story that enters a much broader public discussion about land-use decision-making. Why do you suppose the pickup truck was chosen as a metaphor for self-validating reduction? Do you think it works in this example? In this particular place?

2. Now, how about a more challenging task? Identify a contentious issue in your community that illustrates a potential self-validating reduction. Analyse the language and metaphors used by proponents. Often, at the tip of the iceberg are calls to be reasonable, to be strategic, and to find compromises. These typically come from all angles in an issue. But, the concept of self-validating reduction shows how, in many issues, compromise favours development interests. Once a project is started, however small, there can easily be a tendency for incremental encroachments on wild, social, and hybrid spaces that represent a multitude of values. With all this in mind, write a short essay, an opinion piece, or letter to an editor that is framed by your knowledge of self-validating reduction. Your story can issue warnings, show how compromise is a loaded game to play, or make a thoughtful entry into public discourse. After all, ethics is nothing if it does not engage community members into processes that allow them to assert control over decisions that affect them.

## HANDOUT: PRIVILEGING PROSPECTING, STAKING, AND MINING IS A BREACH OF TRUST!

Yukoners have many dreams about the land. It's tough, cold, and sometimes unforgiving, but we love it. It nourishes our spirit, excites our curiosity, and commands our respect. We dream about many things—having a cabin in the woods, seeing sheep, experiencing the Porcupine caribou herd, catching fish, or just being out in a peaceful corner of this place, filling ourselves with the wonder of it all. Many of us "would trade it for no land on earth."

Our values are reflected in these dreams and, interestingly, in the Yukon Environment Act. The first line recognizes "that the way of life of the people of the Yukon is founded on an economic, cultural, aesthetic and spiritual relationship with the environment and that this relationship is dependent on respect for and protection of the resources of the Yukon." This passage recognizes cultural diversity, and values beyond the economic ones. Yet, it still seems insufficient to offer protection. We are constantly reminded that these values can be pre-empted by other legislation.

The recommended approval of Cash Minerals' Wernecke Winter Road Access Project is a case in point. In spite of unprecedented public opposition, the regional Yukon Environmental and Socio-economic Assessment Board, in Mayo, has given its blessing to this project, and in so doing has put at risk values that the Yukon Environment Act is predicated upon.

Dozens of letters affirmed a broad range of concern about the road project. The letters often did not defer to the technical language of "mitigation," so loved by proponents of technical processes, but they did underscore a broad suite of values. At times, mitigation is not a viable option. Skepticism about reversing industrial damage through mitigation seems befuddling to some mining proponents. The Mayo office has apparently acquiesced to their rhetoric. But when mitigating, wilderness values always lose.

Think of driving a pickup truck out of the showroom and into a parking lot where it gets its first scrape. Would you calmly say, "I can mitigate this?" What would you say about the second scrape, the first dent? And how would you feel after ten years of accumulated scrapes and dents?

There are parallels here. Just as scrapes and dints accumulate in the pickup truck, so, too, do they accumulate in wilderness areas. At first there is outrage, but as they add up over time we take less notice. These are cumulative impacts. At some point, we begin to expect more abuse. When the truck becomes an old "beater" it is easy to treat it roughly. Wilderness advocates know this. They know that one set of scrapes on the land leads to more scrapes and dints. As the land is degraded, there is less respect for it. Mitigation is never a complete success. This is how slow, grinding, incremental change happens.[25]

---

[25] Bob Jickling, long-time Yukon resident, "Privileging prospecting, staking and mining is a breach of trust!"

## Can the cycle of self-validating reduction be broken?

Through leading many workshops on self-validating reduction, we have found that it is an extremely useful analytical tool. It has helped many people gain new perspectives on actions that happen around us. These actions can be slow, incremental, and self-perpetuating. Without this analytical perspective, people can easily lose sight of what is happening, and the potential cumulative risks of so-called "progress". It can be a startling exercise to recognise that a favourite place, someone near to your, or maybe even yourself, has been reduced in this way.

However, if self-validating reduction works like a self-fulfilling prophecy that reduces something in the world, how should we respond? Can this cycle of reduction be countered? What would be the opposite?

Yes, we do believe that this cycle can be countered, and we turn again to Anthony Weston's insight and imagination. He sometimes calls the opposite to self-validating reduction, self-validating invitation; then we are led to question, what would a self-perpetuating invitation look like?

The idea of responding to cultural assumptions and self-validating reductions that we find is exceedingly important and are taken up extensively in Chapter 9 which asks how can we reimagine the future? and in Chapter 10, how elusive is ethical action? We encourage you to look in these chapters for positive responses to the reduction cycles explored here.

## The contested space of resistance

After Donald Trump became president of the United States in 2016, at least one case for educators' visions of education has been put to the test.

The *Huffington Post* reported that teachers at a high school in a predominantly white county of the State of Maryland were asked to take down pro-diversity posters from their classrooms. School administrators issuing the request were responding to complaints from at least one staff member who felt that the posters were "political" and "anti-Trump".[26]

This article quoted a spokeswoman for the county who said, "The school does not allow teachers to put up political posters in their classrooms "unless it's part of a curriculum and they represent both sides."[27]

The reporter goes on to acknowledge that the poster art does recognise groups may feel marginalised under the Trump administration, but that it is not anti-Trump in nature. Furthermore, that this series of posters was not designed to refer to any president or political party.

---

[26] Dana Liebelson, "School Asks Teachers To Take Down Pro-Diversity Posters."
[27] Ibid.

Figure 3.3    Contested resistance, an example of the "We the People" posters the school perceived as "anti-Trump"[28]

## LEARNING ACTIVITY 3.6: LOCATING SPACE FOR RESISTANCE

1. Consider a question from one of the school's students, "Since the posters were taken down, what does that tell the students?" You could also ask if the removal of the posters was inherently pro-Trump?

2. What do you make of the perceived need to "present both sides of the issue?" Does this make sense in complex issues with many dimensions and nuances? Does it make sense to provide equal representation to a view that runs overwhelmingly contrary to the best available evidence? What should you do in an instance like this when contrary examples, often reinforcing the status quo, are already in abundance throughout the school, the curricula, and the community?

3. The students responded by starting an online fundraiser to print free T-shirts, displaying the diversity messages. They planned to wear them

---

28   This poster was accessed from Dana Liebelson, "School Asks Teachers To Take Down Pro-Diversity Posters." Artwork is by Shepard Fairey & Arlene Mejorado.

to stand by those folks represented by the images in the posters. What do you think of this form of resistance? As an educator, can you devise another educational response that is consistent with your vision of education?

## Education and ethics: Implications for teaching and learning

The critique of cultural artefacts and thinking about how to then employ these critiques easily leads to questions about education itself. Questions multiply as interest groups seek to influence curricula. We have touched on this question during the last section of this chapter. We do return to questions of this type again in Chapter 11.

## SOME THOUGHTS ABOUT THEORY

In education, a perennial tension pulls those involved with teaching, learning, and schooling between the twin poles of transmitting cultural values and putting those same values up for debate and revision. In the 21st century it seems more important than ever to revisit this tension. Andrew Sayer, a contemporary social theorist, has noted that "times of major social change prompt far more ethical debate than periods of relative stability."[29] The Earth is stressed with myriad social and environmental problems. The persistent threat of catastrophic climate change is overarching. Doing nothing critical – that is simply transmitting existing values, or seeing them as normal, appropriate, and inevitable – would be to normalise values that are leading us to the brink of disaster. The same would be true of blindly stumbling forward unaware of those values that provide direction and authority to societies where we live. In either event, the outcome is to normalise catastrophe.

A brilliant way of portraying the challenging nature of this educational tension is artistically conveyed in a sculpture that greets visitors to the Kenyan National Museum in Nairobi. A tall tower depicts gourds being tumbled out of a gigantic ladle and cascading downwards. A closer look reveals a smaller number of gourds arching upwards like the fine mist above a waterfall that is blown away in the wind.

---

[29]  Andrew Sayer, *Realism and Social Science.*

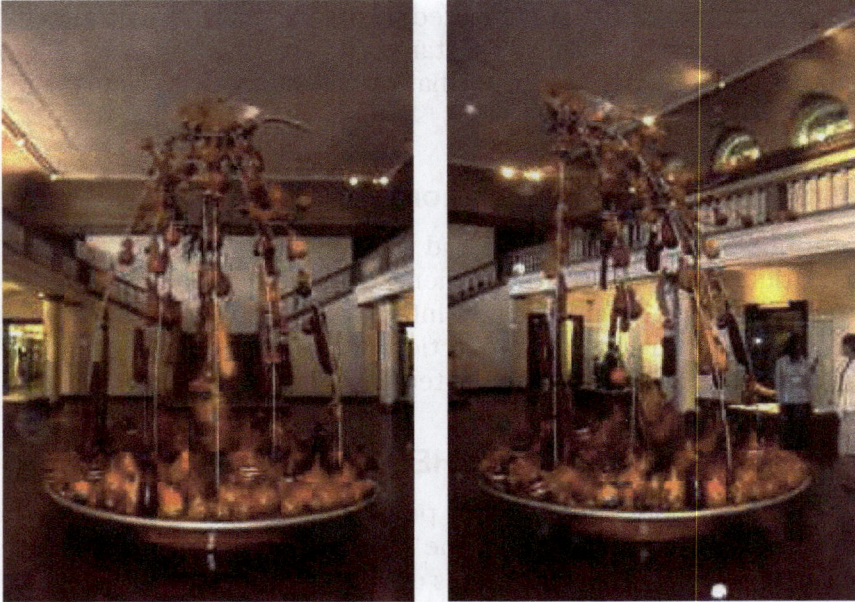

Figure 3.4   Cultural values expressed in gourds
(Photos by author, Bob Jickling)

The gourds themselves are all different, each kind reflecting the particular gourd traditionally used for carrying water by each of Kenya's many tribal groups. The gourds, then, are deeply symbolic of cultural values held by all Kenyans and their passage over this sculpted waterfall depicts the passage of these values from one generation to another. The wisp of mist at the top reminds viewers that some cultural values, however, are not worthy of transmission; they must change or be blown away.

This is not to say that, as authors of this book, we should, or even can, say exactly which values should change, and it is not our educational position that educators should lead students to a set of perceived "right" values. To be honest, we cannot claim to be value-neutral either. The examples we choose to work with do point in directions that we think are worthy of investigating. They help us to explore the practical adequacy of ethics questions. It is our hope that readers will use these examples as starting points only and expand the repertoire of examples – cultural artefacts and issues – to suit the needs of their own students. Still, the first challenge is to gain perspective of the values that are operational in our social contexts. As Neil Evernden reminds us, the authoritative values in a culture can be so deeply embedded in daily life as to be nearly invisible.[30] This, then, is our starting point. In this section we have provided a few activities to help us see values that may be operational in our lives,

---

[30]   Neil Evernden, *The Natural Alien: Humankind and Environment.*

sometimes without our knowledge and consent. Once revealed, we can begin to tackle those values that may be leading us towards catastrophe.

Talking about catastrophe is provocative, we know. Still, important predictions revealed in documents like the Fourth and Fifth Assessment Reports from the Intergovernmental Panel on Climate Change[31] and the Stern Report[32] warn that the future will be much different, and the litany of expected changes is now familiar: higher temperatures, increased drought, fiercer storms, and loss of terrestrial and marine biological diversity, just to name a few. Add to this, as David Orr[33] reminds us, a 30-year thermal lag between the release of heat trapping gasses and climate driven weather events, we can predict a great deal of turmoil in years to come. In early 2016, NASA reported that Earth's surface temperatures in 2015 were the hottest since modern record keeping began in 1880, shattering the previous record set only one year earlier in 2014.[34] According to these and many metrics, we are perilously close to disastrous thresholds.

So, in what ways is this catastrophic? We ask this question seriously, because it takes us to the heart of this project. First, we must be clear that we are not saving the planet. It will get along without humans and has already demonstrated capacity to regenerate complex and interesting life forms and ecosystems, for example, since the dinosaurs. So, this line of questioning is a way of saying that this catastrophe is largely about us – humans and the things we do. The 30-year thermal lag in climate change drives home the responsibility the current crop of folks have for future generations. What humans do now, will be paid for – financially and in suffering – mostly by those who are born today, and those born in years to come. To be sure, the earth and its more-than-human beings will suffer from human actions if things continue as they are now.

So, what happens when we begin to look at persistent social values and their consequences? It seems that this is a first step in considering which values should persist and which should change or perhaps be blown away like the mist above a waterfall. As we think about the consequences of these values and corresponding human behaviours, we individually and collectively are open to several questions. What about those future generations? What responsibility do we have towards them? What about the more-than-human world? How should we live with the rest of creation – however we imagine that term? What is a good way to live? Or, what is a good way to live in a given context? How can we live in ways that benefit the common good? What are good relations between people and societies? And what are good relationships between people and animals, species,

31  IPCC, *Climate Change 2014: Synthesis Report* (Fifth); IPCC, *Climate Change 2007: Synthesis Report* (Fourth).
32  Stern, Nicholas, *The economics of climate change: The Stern review.*
33  See David Orr, *Down to the Wire.*
34  National Aeronautics and Space Administration, "Global climate change: Vital signs of the planet."

ecosystems, or the more-than-human world? In asking these kinds of questions, we are entering the terrain of ethics.

Ethics though, is not just a cerebral exercise. The questions in the previous paragraph are common entry points, but they are not the only way. What we do counts too. Application of our ethical reflections is important. Still, there is more. Consider flipping this process. What if we begin with actions? Can an ethical life grow from what we do? For some ethicists, like Anthony Weston, an important task is to enable an environmental practice and out of that a personal ethic can grow. This approach opens many more questions and possibilities and some of these are explored in this book – for example, Chapter 7 explores the prospect of an ethics-based epistemology. Yet another way of opening performative possibilities is captured by Louise Profeit-LeBlanc.[35] From her Canadian First Nations perspective, ethics are about what we do so that people will tell good stories about us when we are gone.

The work of this chapter, "What are the real authorities in our cultures?" begins with the broadest category, values. The activity, "Being Critical," aims to find and explore some of those values extant in our societies, to make them visible. The activity, "Self-validating reduction" aims to show how existing values can tend to be self-reinforcing and degrading. This is a beginning for the process of ethical inquiry. That is, the deliberate process of examining values, thoughtfully and through active engagement in issues. "Self-validating invitation" is introduced as an approach to begin looking at how the process of reduction and degradation might be reversed. This theme will be taken up again in later chapters on "Ethics in Action, moral proximity, and caring relationships: Going more deeply", Chapter 8, and "How can we reimagine the future?", Chapter 9.

## References

Air Canada. *EnRoute Magazine*, August, 2002.

Bolton, Ken. "Corporate Money Is Seductive, But Do We Want It?" *Yukon News*, March 9, 1994.

Burke, Jeannie and Walker, Eric. *Rock On Yukon*, Whitehorse: Yukon Chamber of Mines, Government of Canada, & Yukon Department of Education, n.d.

Carlsen, Laura. "The People of the corn," *New Internationalist*, December 2, 2004, 12-13, https://newint.org/features/2004/12/01/changing-the-rules-of-agriculture. https://doi.org/10.1055/s-2004-812692

Cushman, John. "Critics Rise Up Against Environmental Education," *New York Times*, April 22, 1997, A10.

---

[35]  Louise Profeit-Leblanc, "Transferring wisdom through storytelling," 15-17.

Department of Basic Education. *Curriculum and assessment policy statement, Grades 10-12: Life Sciences.* Pretoria, Republic of South Africa, 2011, 12.

Eisner, Elliot. *The Educational Imagination: On The Design And Evaluation Of School Programs*, 2nd Edition. New York: MacMillan, 1985.

Evernden, Neil. *The Natural Alien: Humankind and Environment.* Toronto: University of Toronto Press, 1985.

IPCC. *Climate Change 2007: Synthesis Report* (Fourth). Intergovernmental Panel on Climate Change, 2007, accessed June 21, 2019, https://archive .ipcc.ch/report/ar4/syr/. https://doi.org/10.1017/CBO9780511546013

IPCC. *Climate Change 2014: Synthesis Report* (Fifth). Intergovernmental Panel on Climate Change, 2014, accessed June 21, 2019, https://archive .ipcc.ch/report/ar5/syr/

Jickling, Bob. "Making Ethics an Everyday Activity: How Can We Reduce the Barriers?" *Canadian Journal of Environmental Education* 9 (2004): 11-26.

Jickling, Bob. "Privileging prospecting, staking and mining is a breach of trust!" *Yukon News*, January 12, 2008, accessed June 21, 2019, http:// yukon-news.com/news/privileging-prospecting-staking-and-mining-is -a-breach-of-trust/

Liebelson, Dana. "School Asks Teachers To Take Down Pro-Diversity Posters, Saying They're "Anti-Trump," *Huffington Post*, February 22, 2017, accessed June 21, 2019, http://www.huffingtonpost.com/entry /school-pro-diversity-posters-trumpus58ac87b9e4b0e784faa21446

National Aeronautics and Space Administration. "Global climate change: Vital signs of the planet," 2015, accessed June 21, 2019, http://climate .nasa.gov/

Norgaard, Kari Marie. "Climate denial and the construction of innocence: Reproducing Transnational environmental privilege in the face of climate change." *Race, Gender & Class* 19, no. 1/2 (2012): 80-103.

Orr, David. *Down to the Wire: Confronting Climate Collapse.* New York: Oxford University Press, 2009. https://doi.org/10.1093/oso /9780195393538.001.0001

Profeit-Leblanc, Louise. "Transferring wisdom through storytelling," in *A colloquium on environment, ethics, and education*, ed. Bob Jickling. Whitehorse: Yukon College, 1996: 14-19.

Rockström, Johan, Guy Brasseur, Brian Hoskins, Wolfgang Lucht, John Schellnhuber, Pavel Kabat, Nebojsa Nakicenovic, Peng Gong, Peter Schlosser, Maria Máñez Costa, April Humble, Nick Eyre, Peter Gleick, Rachel James, Andre Lucena, Omar Masera, Marcus Moench, Roberto Schaeffer, Sybil Seitzinger, Sander van der Leeuw, BobWard, Nicholas Stern, James Hurrell, Leena Srivastava, Jennifer Morgan, Carlos Nobre, Youba Sokona, Roger Cremades, Ellinor Roth, Diana Liverman, and James Arnott. "Climate change: The necessary, the possible and the desirable Earth League climate statement on the implications for climate policy from the 5th IPCC Assessment." *Earth's Future* 2, no. 12 (2014): 606-611. https://doi.org/10.1002/2014EF000280

Rogers, Katie. "American Hunter Killed Cecil, Beloved Lion Who Was Lured Out of His Sanctuary," *New York Times*, July 28, 2015, accessed June 21, 2019, http://www.nytimes.com/2015/07/29/world/africa/american -hunter-is-accused-of-killing-cecil-a-beloved-lion-in-zimbabwe.html

Roy, Arundhati. *The Algebra of Infinite Justice*. London: Harper Collins, 2002.

Sayer, Andrew. *Realism and Social Science*. London: SAGE, 2000. https://doi .org/10.4135/9781446218730

Schumacher, Ernst Friedrich. *Small Is Beautiful*. New York: Harper & Row, 1973.

Stern, Nicholas. *The economics of climate change: The Stern review*. Cambridge University press, 2007. https://doi.org/10.1017 /CBO9780511817434

South African Airways. World Summit Special, *Sawubona Magazine*, August 2002.

The Guardian. "Black rhino hunt permit brings $350,000 at controversial auction," January 12 2014, accessed June 21, 2019, https://www .theguardian.com/environment/2014/jan/12/black-rhino-hunt-permit -brings-350000-at-controversial-auction

UNESCO. Sustainable Development Goals, accessed October 29, 2020. https://sustainabledevelopment.un.org/topics /sustainabledevelopmentgoals

Weston, Anthony. *Back to earth: Tomorrow's environmentalism*. Philadelphia, PA: Temple University Press, 1994.

Weston, Anthony. "Self-validating Reduction: Toward a Theory of the Devaluation of Nature," *Environmental Ethics* 18, no. 2 (1996): 115-132. https://doi.org/10.5840/enviroethics199618227

# CHAPTER 4

## WHY SHOULD WE CARE ABOUT ANIMALS? NATURE? THE MORE-THAN-HUMAN WORLD?

**Pedagogical intent:** Ethics questions are often more complex than they initially seem. Looking critically at everyday experiences can help us to make more ethically inspired choices in our daily lives. This chapter introduces teachers to more systemic ways of thinking in which humans and more-than-humans can be seen as one interrelated system rather than as separate entities with one being more powerful than the other or with one being exploitable in the interests of the other.

**Activity:**                    4.1: Exploring stories with no easy answer

**Main theoretical resources:** In our experience, quandaries involving animals and other more-than-human beings often invoke discussions about animal rights and ecosystem welfare. With this in mind, we have drawn on Peter Singer and his work on animal liberation, Tom Regan and animal rights theorising, and Paul Taylor's broadly inclusive environmental ethic grounded in what he calls "respect for nature". These theoretical introductions can be useful in deepening discussion.

## Complex questions and ethical quandaries

### Why quandaries?

First think about dilemmas. We sometimes talk about "being on the horns of a dilemma" when we have a difficult judgement to make between two choices. However, environmental issues seldom boil down to only two choices. They are more often complex, multifaceted, and reflect multiple perspectives. To get away from the idea that issues can be reduced to two perspectives, we like to use the word "quandary" to describe complex and perplexing ethics questions.

There are many stories that illustrate quandaries. Here we discuss one such story in some depth – The Great Elephant Debate – and invite you to investigate other similar stories.

### The Great Elephant Debate

A visit to any of the big national parks in South Africa, Botswana, Zambia, and Namibia tell a similar story. One is likely to meet hundreds of elephants. They are truly majestic and observing them can only engender enormous respect and awe. Elephants are sensitive animals. They have complex social structures and can communicate fear and pain to other elephants over huge distances.

During such travels, one is likely to come across game rangers and community members saying something like this: "There are too many elephants", but how do we establish whether there are "too many elephants"? What does this mean for people, for the elephants, for biodiversity?

Since protection for African elephants was instituted in the 1980s (through CITES,[1] anti-poaching interventions, and wildlife management), elephants have multiplied in the southern African region. Today large tracts of land seem to be irreversibly changed by elephants and their activities (it reminds one a little of human beings, who have also multiplied, and the way they are able to change the landscape). While there is no proof that elephants reduce biodiversity, some critics comment on a scarcity of small mammals and birds in places where elephants were abundant. Others say that elephants can assist with biodiversity through seed dispersal and generation of habitat, especially for birds. Elephants nowadays are "contained" by boundaries of national parks while, in earlier times, they would have roamed freely across the African plains on their annual migrations. To complicate the story, national parks are important sources of income for the local economies as they attract many tourists. Poverty is a profound challenge in southern African countries and industries such as tourism are increasingly important responses. Tourism generates jobs and provides outlets for craft sales and local activities. Conservation officials and scientists who manage parks aim to maximise both tourism and the biodiversity potential.

In South Africa, elephant culling was stopped in 1995 because of international pressure. Since then, the elephant population in the Kruger National Park has grown from 7 000 to at around 6.5% annually, reaching around 16 900 during the 2012 count.[2] These elephants were reproducing at about 1 000 per year, and each elephant eats up to 150 kg of vegetation per day. So, a Great Elephant Debate is taking place in South Africa around the question of what constitutes "too many elephants". It is a complex debate, with few easy solutions. Kruger National Park has developed an elephant management strategy, based on concerns raised by conservation managers about perceived threats to biodiversity in the Park – resulting from "too many elephants". The reality is that those conservation strategies, and other human activities, make the original experience of elephants to roam free and independent of human intervention, impossible. South African National Parks (SAN Parks) are aware of the different voices in the debate and are acutely aware that they are left with imperfect situations and

---

[1]    CITES: Convention International Trade in Endangered Species of wild fauna and flora.

[2]    Ann Toon and Steve Toon, "A Conservationists' Conundrum."

solutions. This debate is so complex that the Minister of Environment and Tourism consulted international partners for advice on how South Africa should address the biodiversity questions raised by the Great Elephant Debate in the region.

## An ethical quandary

SAN Parks are faced with difficult choices. What are they to do about the burgeoning elephant populations in parks? Animal rights groups uphold intrinsic principles concerning rights of animals to live unhampered by human intervention. Sustainable use groups argue for economic aspects of elephant management, with an emphasis on how local communities could benefit from hunting and/or from selling products from culled elephants. Conservationists and scientists maintain that the ecological balance in the park is most important. However, there is little certainty as to exactly what this balance is, and how it ought to be established. The debate is complicated by serious economic and social implications associated with the costs of maintaining a contraceptive programme, culling, or constantly expanding parks – involving the relocation of human populations – to accommodate growing numbers of elephants.[3] SAN Parks management and other government leaders recognise the complexity of the issue and quandary they face; they note that it would be naïve to expect a perfect solution.

## Trying alternatives

In response to the complex ethical questions, SAN Parks have explored alternatives to culling. They have tried contraceptive methods and disrupting social relationships amongst the elephants and the second is too expensive for a developing country. "Trans-frontier" Parks have been established to create bigger parks and migration routes for the elephants. Translocation of elephants has taken place, but this too, is an expensive process.

A question asked in the newspapers is whether culling is a viable management strategy. Given the complexity of the question, SAN Parks turned to the public and hosted the "Great Elephant Debate" in 2005. In a talk on the ethics of elephant management, Saliem Fakir,[4] IUCN Country Programme Coordinator, said we need more adequate language for deliberating the elephant issue. He argues for *tolerance*, for *ethical pluralism* (a mix of different ethical viewpoints), for locating discussions

---

3    Fred Brigland, "5000 elephants must die. Here's why"; South African Government, "Elephant Debate Not a Ploy to Reintroduce Culling"; Richard Leakey and Virginia Morell, *Wildlife Wars*; Anna Whitehouse and Pat Irwin, *A field guide to the Addo elephants*; Sharon Hammond, "Kruger won't cull elephants"; and WEESA, "Great Elephant Debate."

4    Saliem Fakir, "Notes on the Ethics of Elephant Culling."

in *context*, and for considering technological and economic dimensions of the issue. He warns, for example, that "sustainable use" arguments are confused with ecological questions, and economic questions are side-tracked by commercial interests. He argues for "sound judgement" based on available evidence, careful deliberation, and consideration of long and short-term trade-offs. He urges citizens to build a *listening* and a *learning* society.

# HANDOUT: SOME THOUGHTS ON FINDING ETHICAL PRACTICES IN DIFFICULT ISSUES

International literature is replete with references to the need for a "new ethic" (see for example UNESCO 2015,[5] the Earth Charter Initiative[6]). There is much international support for framing this new ethic in terms of "sustainability." Will this assist South African National Parks in making their choices to resolve the Great Elephant Debate? Is the solution that simple? How, for example, would it respond if the ethic of sustainable use were "skewed" towards commercial interests in elephants? And just what does "sustainability" really mean in this context? Fewer elephants? Bigger parks? More effective contraception for elephants? Trophy hunting to generate income for communities? Relocation of rural communities away from the boundaries of the park to enable continued expansion? More money from the international community to pay for expensive alternatives?

Ethical choices cannot be easily defined. The process of engaging ethical questions is often a difficult and time-consuming process involving deliberations, trade-offs, development of creative alternatives, and considering realities in different contexts. This is acknowledged. From the elephant story, it would seem that ethical questions are often *ambivalent* and that there are no simple recipes, and no "simple ethics."

In looking more deeply into the elephant story and other such stories, some "new lines of thinking" may be opened up for our ethical work in the everyday:

Ethical practice often encounters *ambivalence* and *ambiguity*.

It involves finding ways to allow people to *deliberate* on ethical questions, and to engage critically with ethical quandaries—in diverse contexts. This process requires *respect* and *tolerance*, to hear and consider different perspectives.

When we are tolerant and respectful we find ways to share common language and create space to discuss alternatives.

This process, however, is mediated by an "*ethic of timeliness.*" According to Johan Hattingh[7] an "ethic of timeliness" is an imperative to take action before reaching the point of no return, before outcomes become irreversible. This often requires a combination of pragmatic decision-making, creativity and a commitment to exploring all available alternatives within time boundaries.

*Complexity* is also an ethical issue – the simplification of complex issues (such as the Great Elephant Debate) is both fraudulent and irresponsible. This places responsibility on the shoulders of everyone, including the scientific and intellectual communities – and especially educators – to work closely with decision makers, the public, and each other so that all are aware of the multiple dimensions of issues and associated risks.

---

[5]   UNESCO, "Education 2030 Framework for Action."
[6]   See Earth Charter Initiative, "Earth Charter Around the world."
[7]   Johan Hattingh, Unpublished M.Ed course notes.

# LEARNING ACTIVITY 4.1: EXPLORING STORIES WITH NO EASY ANSWER

Taking a close look at the quandary generated by the Great Elephant Debate can provide opportunities to examine personal and cultural beliefs, consider ethics in practice, and have ideas examined and discussed by peers. With this in mind, consider the following activity:

1. Gather into small groups to discuss the Great Elephant Debate. What would you do? Why? What assumptions do you hold about human-elephant relationships? Share your thoughts. Can your group agree on a collective response? What else do you need to know to take your discussions further?

2. How can an "ethic of timeliness", as well as increased understanding of complexities, change the deliberations? See the handout above.

3. Some suggest that when we come up against difficult questions or quandaries, we can sometimes shift the problems by reframing the questions. Are there other ways of looking at elephant issues that can help us transform the problem, to come at it from a different angle?

## Reframing problems

Some researchers have begun trying to re-frame the issue and shift the problems in elephant-human relationships. These do not always "solve the problems", but they have been successful in creating alternatives that local people can work with. Some examples include:

- The establishment of Trans-frontier Parks that allow elephants to move along larger migration routes. This allows for biodiversity regeneration as the elephants move to new places.
- Not putting up park fences but rather supporting local farmers to fence their crops so that people and elephants can live alongside each other. In Zambia, the Department of Wildlife have created "zones" outside of the park boundaries called Game Management Areas where people and elephants live side-by-side. The challenge has, however, been to provide local people with the support for fencing their crops.
- We could also think of reading the story "the other way round" i.e., from the view of the elephants. How are the elephants changing people? What might the elephants be thinking about people having a Great Elephant Debate?

Solutions to reframing a problem have not been easy, especially as elephants sometimes threaten the lives of people in the communities, and authorities are called upon to address the elephant problem by shooting one or more of the elephants. Community-animal relationships are not always harmonious. But the challenge here is to find new ways to work with the issue and to reduce conflict.[8]

## EXTENSION ACTIVITY

There are many other stories that also have "no easy answers". One example that comes to mind is the contemporary use of DDT (Dichlorodiphenyltrichloroethane), in some places, to prevent the spread of malaria in rural Africa (more people in rural Africa currently die from Malaria than any other disease). The use of DDT is apparently a "proven" solution to the problem, but what else needs to be considered? Some other "stories with no easy answers" may include: The building of large dams (e.g., the Nujiang dam in China or the Narmada dam in India – see www .dams.org), or the current trend to privatise water in Africa, Latin America and Asia.

4. Find out about one of the above-mentioned issues, or study a local quandary that appears to have "no easy answers". Seek contrasting perspectives and use materials from different sources. The internet normally provides a good starting point. Magazines such as the *New Internationalist* (www.newinte.org), the *National Geographic* (nationalgeographic.com), and the *New Scientist* (www.newscientist .com) provide good quality sources with different perspectives. Try locally available magazines or newspapers too.

   What are the key issues and what are the difficult questions and the ethical quandaries? What choices are available? What kinds of deliberation processes may be needed to resolve the quandaries? Who should be involved in the deliberations? What creative alternatives are possible to resolve and reframe the questions? Are ethics framed by the concept of "sustainability" adequate for resolving the quandaries you are working on?

5. Examine the Earth Charter at www.earthcharter.org, or another "ethical framework". How can this framework and its listed principles help stakeholders to deliberate on ethical alternatives to the Great Elephant Debate or other issues? What is the value of this approach, and what are its limitations?

The Earth Charter and further options for using it will be discussed with more detail in Chapter 11.

---

[8]   See for example, Anthony Weston, A 21st *Century Ethical Toolbox*, 190. For a further discussion of these issues and "reframing" examples see also, Suzanne Hamel, *A participatory approach to community-based curriculum development*. For a literary interpretation of events from the perspective of elephants see Barbara Gowdy, *The White Bone*.

# SOME THOUGHTS ABOUT THEORY

## Ethics theories from animal stories

Relationships between Indigenous societies and the rest of creation have been mediated through stories for a long time, probably since the very first stories were told. Western societies and other societies strongly influenced by these early stories, often seem to have lost track of the intimate relationships with their landscapes and places that would have been embedded in original stories. As a whole they have lost track of their Indigeneity and the stories that were once a source of guidance in navigating the complexities of living well in a place. This loss has been occurring for a long time.

The so-called "Enlightenment" period is one popular marker for an acceleration of the transition into a "modern society". While events of this era can be credited with easing the influence of religious dogma from conceptions of knowledge and everyday life, there have been costs. Intimacy, feelings, care – anything that might be considered subjective, in today's terms – have been devalued. The resulting action-guiding stories can be framed in many ways – scientism, capitalism, colonialism, corporatism, globalism, neo-liberalism, neo-conservatism, the list goes on. In response to these emerging stories, ethicists have been inventing new ways to mediate relationships between each other and amongst societies. More recently the more-than-human world has joined this circle of concern. Hence, the field of environmental ethics was born. The work of these ethicists is typically framed as theoretical. Though, it can also be said that the form that ethics ultimately takes is determined by how ethics are used, by what people do.[9]

The quandaries present in the activities of this chapter are both exercises in seeing how ethics are playing out through contemporary examples, and imaginings about how ethics could be implemented. They are exercises in thinking about the work that ethics can do in the world.

In conducting these kinds of activities in workshops and classes, some touchstones arise. Broadly speaking, these are typically concerns about animal welfare. More specifically, there are common references to the "rights" of both animals and people and concerns about animal suffering. Added to this combination of more intellectualised positions, there is often a more visceral sympathy for the beings involved, both human and more-than-human. We will return to this more visceral component later. First, we will explore these links to rights and suffering and some of the theoretical underpinnings by looking at the three authors Peter Singer, Tom Regan, and Paul Taylor.

---

9    Eugene Hargrove, "Science, ethics, and the care of ecosystems," 44-61.

## Extensionism

The philosopher Peter Singer ignited concern for animals in 1975 with his landmark book, *Animal Liberation*.[10] In making his case for animals, he evoked other liberation struggles. For example, the struggle for liberation from slavery was grounded on the principle of equality between races. Similarly, women's issues, framed as a liberation movement in his time, were, and continue to be a struggle against many existing inequalities between sexes (and/or genders in current discourses). Following utilitarian philosopher Jeremy Bentham, he argued that any being that has interests should have those interests taken into account equally. For Singer, justice demands equality in standing and treatment. To deny economic and social equality based on race and sex were, for him, racist and sexist.

With historical liberation movements as background, Singer examined the treatment of animals. Again, following Bentham, he asked if animals feel pain and if they suffer, he argued that they surely have interests in minimising suffering. It follows, then, that if a being suffers then its interests in pursuit of happiness and avoiding pain must be taken into consideration. For Singer, the ability to feel pain – sometimes referred to as sentience – is the basic moral criterion.

At the time, Singer cautiously acknowledged that we only infer that animals feel pain – through observations of writhing, yelping, and avoidance of the source of pain. Today there appears to be little doubt that this inference is correct. Few doubt that animals do, indeed, experience pain. For Singer, if animals feel pain, there can be no moral justification for ignoring their suffering. Treating other species in ways that would be wrong to treat our own would be "speciesism". Singer's project, then, has been to liberate other beings from the tyranny of speciesism.

Singer has used his theory of animal liberation to problematise factory farming where animals are kept in conditions that create enormous pain and suffering to produce inexpensive meat for human consumption. He has also taken on animal testing and argued against the gratuitous pain endured by animals, especially in testing cosmetics and household products.

When asked about more difficult topics – such as of medical research, or questions about what to do about rats biting children living in slums – Singer acknowledges that there are difficult instances of genuine conflicts of interest. The essential point is, however, that these are first, conflict of interest, and that we recognise that rats have interests. Of course the quandaries presented in this chapter describe many conflicts of interests, too. However, they invite readers to consider those interests kept at the margins of public discussions, or simply overlooked. How, for example,

---

10   See Peter Singer, *Animal Liberation*.

can the interests of elephants be taken seriously when they conflict with human interests?

Singer's theorising has often been described as an early example of moral extensionism in environmental ethics. This means that existing moral frameworks have been extended to include the interests of entities typically discounted in ethical contexts. Here two things have happened. One, the logic of liberation movements has been extended to include species other than humans, and two, the utilitarian framework developed by Bentham has been extended to include the interests of the more-than-human world. There are, of course, other ways to extend moral theorizing and Tom Regan's theorizing about animal rights is taken up next.

\*\*\*

Tom Regan, in an equally riveting foray into moral extensionism, begins with a different premise than Singer.[11] For him, the fundamental wrongs are social and intellectual systems commonly present in the 20th and 21st centuries that allow us to see animals as resources for human benefit. In Regan's view it is wrong that we can treat animals as utilities, as means to human ends, as instrumentally valuable without regard to their own inherent or intrinsic value.

Regan looks to similarities between animals and humans as a basis for intrinsic value. For him, the bedrock similarity is that we are all conscious creatures, aware of our individual well-being, regardless of our utility to somebody else. What animals and humans hold in common is that we are, thus, subjects-of-a-life; we are experiencing, goal-directed beings, conscious of our own welfare. Of course, one imposing question is, where do we draw the line for subjects-of-a-life? Are grizzly bears in and oysters out? For Regan, clear cases are mammals over the age of one year; all such subjects-of-a-life are rights holders. Further, for him, all those who hold rights do so equally. Regan's approach here is an example of Kant's concept of a categorical imperative, where an ultimate principle is identified for a given situation, and the same principle is applied equally across all similar situations.[12]

For Regan, the fundamental principle is that subjects-of-a-life are rights holders and from this principle several goals are indicated including:
- the total abolition of the use of animals in science,
- the total dissolution of commercial animal agriculture, and
- the total elimination of sport hunting and trapping.[13]

---

[11] Tom Regan, "The case for animal rights."
[12] Kenneth Strike and Jonas Soltis, *The ethics of teaching*.
[13] Tom Regan, "The case for animal rights," 13-26.

For Regan it is not possible to change unjust institutions by tidying them up – by condemning some animal uses while rationalising others. For him, animals must be extended the same moral consideration as humans.

At this point, it is worth considering how some animal rights perspectives differ from animal liberationists. First, as Regan puts it,[14] animal suffering is wrong, but it is not the most fundamental wrong. Abuse of animals, causing them to live in stressful and often painful conditions is, for him, just a symptom of deeper and more systemic wrong. The point here is not to just minimise pain and suffering, but to cease treating animals as a resource for humans, as a means to our human ends. Second, it is also possible to argue that the utilitarian position of seeking to minimise pain and suffering in the interests of maximising happiness, can override intrinsic value and rights of individuals in the interest of a greater common good.

A critical reader will also see potential weaknesses in both theories, and some of these will be taken up later. But, for now, we would like to think about the limits of animal rights theorising as a basis for an environmental ethic. How, for example, is such a theory adequate when so much of the more-than-human world is left out?

In fairness to Regan, he explicitly left room for further theoretical development.[15] For him, the criterion of subject-of-a-life was sufficient to acknowledge inherent value. He also speculated that other non-conscious more-than-humans might also be inherently valuable, although he was unclear about how this might be theorised. Paul Taylor[16] is one theorist who took up the challenge to extend this line of theorising. His work is the subject of the next section.

\*\*\*

Like Singer and Regan, Paul Taylor sought to extend moral consideration beyond the realm of humans. Like the other two, he also sought a principled basis for his system of ethics – one that was grounded in rationality and could be applied universally. What distinguishes his approach is that it does not rely on foundations as limited as sentience or self-awareness, neither of which seems suitable as a basis for a broadly inclusive environmental ethic. Rather, Taylor advances a life-centred system or, as it has come to be more commonly termed, a bio-centric outlook grounded in his moral attitude towards nature – or principle – that he calls "respect for nature". From this standpoint, he argues that a human-centred system of ethics – as something concerned only with duties to humans – fails to protect the good of more-than-human beings or, in other words, their inherent or intrinsic value.[17]

---

14  Ibid.
15  Tom Regan, "The case for animal rights."
16  Paul Taylor, *Respect for Nature.*
17  Ibid.

The line of reasoning leading to intrinsic value being recognised in wild plants and animals – regardless of their level of sentience or self-awareness – begins in the realisation that all organisms do those things that tend to preserve their life and well-being. In other words, as biological entities, they pursue their own good in their own way. Trees grow towards sunlight, mother bears protect their young, and spiders weave amazing webs to catch prey for their sustenance. Another way of saying this is that each organism is a teleological centre of life, from the Greek *telos*, meaning end or purpose. Their tendencies are ends in themselves, intrinsic values. Our duties then are to protect or promote more-than-human organisms' good for their own sake.

From his recognition that organisms are intrinsically valuable, Taylor has derived four rules to guide actions. Put simply these are:

- *The Rule of Non-maleficence* is the duty to do no harm – to organisms, species, or a biotic community.
- *The Rule of Non-interference* represents duties to refrain from restricting the freedom of individual organisms and disrupting the functioning of ecosystems and biotic communities.
- *The Rule of Fidelity* is the duty that humans have to not mislead or break a trust that wild animals may have acquired in humans, based on past actions. Deception with the intent to harm is the kind of thing covered here.
- *The Rule of Restitutive Justice* is the duty to make amends to moral subjects for instances of injustice by some form of compensation or reparation.

In reading these rules, it is easy to see that they represent a broad and deep respect for living beings. The fourth rule allows that there can and will be instances of serious conflict between the duties of human ethics and environmental ethics. However, there is also something that we can do about this; we can give something back as a measure of respect. Of course, this is not really a new idea. Traditional cultures continue to base their daily practices on ideas of respect enacted through daily activities, notably through acts of thanks and gift-giving to the entities that sacrificed themselves for human needs.

Some critics might say that if theories of animal liberation and animal rights do not go far enough to provide a basis for an environmental ethic, Taylor's respect for nature goes too far. Its sweep is too broad; it brings humans into far too much conflict with interests of biological entities, such as trees and swamps. Others might worry that the focus on individual biological entities discounts concerns for endangered species and whole ecosystems. On this point, Holmes Rolston III has provided an apropos and entertaining view of species and ecosystems as similarly self-directed.[18]

---

[18] Holmes Rolston, "Ethics on the home planet," 107-139.

## Why should we consider – even study – these theories?

From the descriptions above, it seems that arguably all the positions discussed are theoretically flawed. They might be considered too narrowly conceived to provide a basis for an environmental ethic or, on the other hand, too broad in scope. Considerations of serious conflicts of interests are always a challenge. Singer acknowledges this and suggests a starting point – that we at least recognise that all sentient beings have interests. It could be argued that Regan is coy. Amongst his goals is the total elimination of sport hunting and trapping. Notable is his omission of discussion about subsistence hunting and trapping. It is difficult to fully subscribe to animal rights theorising in a landscape, like the arctic, where vegetables cannot be grown in subsistence quantities. Here, there is a serious theoretical breakdown. Political tension and conflict – between traditional hunters and trappers and anti-fur activists, for example – often arises without offering any path for resolution of these conflicts. The theories are undermined, so why should we spend so much time studying them?

There are some good answers to this question. The first of these arises from our experiences doing environmental ethics workshops. Often it seems that when participants are presented with a quandary, such as the elephant culling debate in South Africa, there is a natural affinity for animals, their welfare, their suffering, and their rights. It also seems that they quickly learn about the difficulties in resolving conflicting interests.

The important thing here is that the work of these theorists can open new ideas and questions. Should we consider animals in their own right? Their welfare? Their suffering? Their rights? If so, how? What is a good way to live with our more-than-human brothers and sisters, all our relations? Do they deserve our moral consideration? Especially when we think about topical issues like elephant culling, killing dolphins, and poaching rhinoceroses. Through questions like these, ethics can open possibilities for people to see the world in different ways. Through the rationales given for these theories, we can begin to acquire fragments of ideas that we can work with. It may not be clear how a conflict of interest can be resolved, but at the same time, it may not feel right to treat animals simply as a means to achieve human ends. Or, in the face of some ethically offensive act, it might be useful to ask what kind of retribution can be returned – to a species or an ecosystem?

In societies dominated by preferences for objective knowledge, facts, and measurable outcomes, it can even be difficult to find language in their mainstream discourses to talk about ethics. Anecdotally, at least, the vocabulary of ethics is becoming more present. Look out for this when doing some of the quandary exercises. We believe we see words like rights, sentience, welfare, intrinsic value, and bio-centric values used more frequently. Other innovations in language use, like "more-than-human" and "self-validating-reduction" can also serve to disrupt the sometimes too comfortable flow of daily discourse.

### If these ideas are flawed as theories, what happens if we treat them as stories?

We discussed the work of stories in Chapter 2. This leads us to wonder what would happen if we stopped thinking of the theories discussed in this chapter as codes, operating rules, fundamental principles, duties, moral obligations, and categorical imperatives. In our wondering, we have come to think that we could also treat these theories as stories. The key thing here is that we do not judge them for their truth – values, objectiveness, logical coherence, correctness, capacity for universal application, correctness, or their ability to evaluate and judge. What if we thought about them as stories with the capacity to do work?

In doing work, we are expecting stories to help us to do some of the following: see issues critically, uncover unseen perspectives, reframe problems, weigh contesting positions, imagine alternatives, be in the world differently, or even bring new language and metaphors to bear. We expect readers will be able to add further possibilities to this list. In other words, while theoretical perspectives may be flawed, they may still serve, when used with care, as action-guiding stories. To the degree that stories enable action, they respect the notion of an *ethic of timeliness* discussed earlier, in that they help people move from ethical deliberation into ethical action, even tentatively.

Framed as action-guiding stories, it seems that they have done a lot of work since the mid-1970s when they began to burst onto the scene. Vegetarianism has, in some places, gone from a fringe activity to mainstream. Standards for farming in many countries have vastly improved, as have standards for research experimentation with animals. India has made environmental news by attributing "non-human personhood" status for cetaceans – dolphins, whales, porpoises – and banning dolphinariums and dolphin shows. While these examples fall far short of the categorical imperatives of Regan, they do suggest that some ethical progress has been made – that these ethical stories have done some good work. As stories, they have the capacity to continue to bring ethical deliberations to bear on everyday decisions. These small ethical invitations show potential for self-validation, leading to ever-expanding ethical considerations; if dolphins and whales, maybe also sharks and wolves can be thought of as persons!

### When cultures clash

Sometimes theorising about ethics that arises in one part of the world can come into conflict with traditional cultural values from other parts of the world. In our experience, the results can range from thoughtful engagement with this dissonance to strong resistance. In order to field tensions and clashes that may arise in educational settings, it is important to highlight that these theories are making propositions about ways of seeing the world but are not imposing or prescribing how we should act in the world. Ethics is a process of research and exploration of ways of being rather than a set of rules that one person or culture can dictate to another.

As educators, we have struggled with question. So far, we have found no clear or definitive answer. Throughout the writing of this book, we have paused to reflect on the questions of cultural dissonance and some of our responses to particular incidents. Importantly, the question posed will endure for all who care about both ethics and cultural issues.

For example, a particularly challenging example arose in our own teaching while presenting an introductory session on extensionist ethics in a Southern African context. A documentary about Peter Singer, his work, his life, and applications of this theorising did challenge cultural values during an environmental ethics seminar in South Africa. Here, particularly graphic images of animals suffering on factory farms and in product testing were riveting to the viewers, and disturbing. Questions following the screening were particularly animated and persistently probing. The film had clearly generated a lot of internal dissonance.

Eventually, it was revealed, in a passionate critique of the video, that Singer's theorising was in direct opposition to deeply held cultural values. In this case, minimising pain and suffering conflicts with the belief that ancestors are guided home by the cries of animals during their slaughter. In this example, causing pain and animal cries, is a virtue for people who share this cultural understanding. It was clear that these opposing values created a quandary for some participants. One of many important questions that followed was, "What are we to do?"

Like all tough questions, there is seldom a single answer for all; answers might change over time. In this instance, as in any quandary, there are genuine conflicts of interest at play. Our answer at the time was to position the ethical theories as stories with the capacity to do work. As theories, they would be highly prescriptive, and justifiably worrisome, but as stories, we can return the question, with another question: "If they do not provide clear answers, can they at least do work?", "Can they help to make some kind of progress in ethical thinking and practice?", "What would progress look like?", and alternatively, "Can the quandary help to reframe important questions that can then help to get work done?"

Reframing questions, or showing new ways of seeing and being, can lead to changes in thinking where old ideas might blow away like the wisps of mist rising from the gourd sculpture, as described in Chapter 3. Or they might lead to previously unimagined new practices. Ultimately, however, each person or group would need to decide on their own way forward – be it a change or a reproduction in the status quo – based on their own needs, context, and culture.

Interestingly, at about the same time as the classroom experience described above, an internet blog posting reflected a similar struggle. While there is no explicit reference to the theorising presented in this chapter, the post does suggest that similar thinking is present in everyday contexts and does have some capacity to do work and maybe even "move the ball down the field" a bit. Xola, a blogger, said:

> I grew up in very traditional ways and slaughtering of cows, goats, chickens, and many other animals [and that] never really moved me emotionally because that practice has become part of who I am. My heart sank today when I had to watch a poached Rhino being treated face to face. I saw that beast lying there looking very unnatural because its horn has been removed in the most inhumane way. Poaching is a brutal exercise! What is interesting for me though is that those are the exact emotions I feel every time I go to eThembeni, Zolani, Phaphamani, Hoorgenoeg, Vergenoeg and many other marginalised communities. Think about going to a bush every time you need a toilet. In a civilised society this is humiliating and it strips one's dignity off, it is brutal in many ways. There are many other things but I'll stop there. Maybe I feel like this because I experienced it first-hand and I still see many people, in fact the majority of South Africans, living in this way. These are the same people who are used by the multi-nationals and high profile politicians to perpetuate this act. This is the majority that can't relate with these animals because they don't own them and never interact with them. After all these people were removed from their homes, which were food producing farms, and the farms were changed to game reserves. Poverty, greed, corruption, capitalism are all responsible for all the injustices in the world.
>
> The injustice of Rhino poaching is but part of a bigger problem. The problem is systematic and structural. My heart is bleeding for the injustices done to the animals, environment and the people of the world. Ours should be a revolution for respect for humans, animals and plant life alike.[19]

These are just two examples of how ideas – related explicitly, implicitly, or just resonant with extensionist ethics – can open up conversations and reflections. They are of a kind that we can call ethical, though it is clearly not required that the word ethics be used. Seen this way, ethics is just part of a larger story, and if it is a good one told well, it can do useful work.

---

19   Reproduced with permission from blogger, Xola Mali. The original blog post no longer appears available on the internet.

## Limits of deontological ethics

Often ethical positions and practices are guided by a desire to adhere to a principle or set of rules, or by a sense of duty or obligation. Such an ethical position is termed "deontological" from the Greek word *deon*, or duty.

Arne Næss observes that human capacity for loving, based on moral duties or compliance with codes, is extremely limited. Often in such circumstances we act against our inclination but comply out of respect for the moral law. His alternative is to seek ways to do what is right because of positive inclinations – out of joy, for example. When we do so, we perform a beautiful act.[20]

John Livingston[21] and Zygmunt Bauman[22] are similarly concerned about deontological ethics and resulting moral codes. For Livingston, they are unknown in nature and, as human creations, are more like prosthetic devices. For him it is important to develop an extended consciousness beyond the mere self. Bauman argues that complying with moral codes reduces responsibilities for one's own moral actions; codes actively erode our moral impulses. For Næss and Bauman, it is more important to find ways to develop people's inclinations, or moral impulses, than to develop their morals.

Concerns about deontological ethics lead directly to work taken up in Chapter 5. They are also taken up in much of the subsequent work in this book, particularly in Chapters 6 and 8.

## Limits to rationalist ethics

One striking feature of Singer, Regan, and Taylor's ethics is their dependence on rationality for theory making. Peter Singer has eschewed sentimental appeals for sympathy towards animals. Indeed, he has often claimed that he is not an animal lover. For him, opposition to injustice is not achieved through emotional appeal, but through careful appeal to moral principles that he assumes we all accept.[23]

Similarly, both Regan and Taylor commit themselves exclusively to rational inquiry for establishing moral principles to guide treatment of animals and treatment of natural ecosystems and wild communities. Regan, echoing Singer, says we must make "a concerted effort not to indulge our emotions or parade our sentiments."[24]

So, what should we do with our sentiments? Can we even have an ethics without some kind of emotional engagement? What about rationality; can

---

20  See Arne Næss, "Self realization: An ecological approach," 19-30; and Arne Næss, *Life's Philosophy*.
21  John Livingston, *Rogue primate*.
22  Zygmunt Bauman, *Postmodern ethics*.
23  See Peter Singer, *Animal Liberation*.
24  Tom Regan, *The case for animal rights*, xii.

it really be free of feelings? These questions point to some more limitations to rationalist ethics as presented here.

Early eco-feminists reject the objective, disinterested stance of rationalist theorising. For Karen Warren,[25] key values – the twin dominations of women and nature, for example – are rooted in historical and socio-economic circumstances. Values of care, love, friendship, and trust are central to understanding who we are in the world. Val Plumwood argues that there "are moral feelings, but they involve reason, behaviour and emotion in ways that do not seem separable."[26]

Regan defends the major conclusions of his book because they are supported by the best arguments. It is important, for him, that animal rights has reason on its side, not just emotion.[27] He acknowledged that his work is cerebral and not the makings for deep passion. However, he did open space for emotion in his final assessment. For him, philosophy should not be too cerebral, but rather a manifestation of disciplined passion.

Concerns and questions raised here will be taken up in explicitly Chapter 6, but also throughout the balance of the book.

## References

Bauman, Zygmunt. *Postmodern ethics*. Oxford: Blackwell Publishers, 1993.

Brigland, Fred. "5000 elephants must die. Here's why," Sunday *Herald*, October 24, 2004, accessed July 10, 2019, http://www.coldtype.net /Assets.04/Voices.04/Voices182.04.pdf

Earth Charter Initiative. "Earth Charter Around the world," accessed July 10, 2019, https://earthcharter.org

Fakir, Saliem. "Notes on the Ethics of Elephant Culling." Talk at the Ethics Society Congress of South Africa, 30 March 2004. Pretoria, South Africa.

Gowdy, Barbara. *The White Bone*. Toronto, Ontario: Harper Collins, 1998.

Hamel, Suzanne. A participatory approach to community-based curriculum development for the Living With Elephants Outreach Program in Botswana. Unpublished M.Ed thesis, Faculty of Education, Lakehead University, 2004.

Hammond, Sharon. "Kruger won't cull elephants," News24, May 27, 2004, accessed July 10, 2019. https://www.news24.com/Africa/News/Kruger -wont-cull-elephants-20040527

Hargrove, Eugene. "Science, ethics, and the care of ecosystems," in *Northern protected areas and wilderness*, eds. Juri Peepre and Bob Jickling. Whitehorse: Canadian Parks and Wilderness Society, Yukon Chapter & Yukon College, 1994: 44-61

---

[25] Karen Warren, "The power and the promise of ecological feminism," 125-46.
[26] Val Plumwood, "Nature, Self, and Gender," 9.
[27] Tom Regan, "The case for animal rights."

Hattingh, Johan. Unpublished M.Ed course notes, Rhodes University, 1999.

Leakey, Richard. & Morell, Virginia. *Wildlife Wars. My battle to save Kenya's elephants*. New York: Pan Books, 2002.

Livingston, John. *Rogue primate: An exploration of human domestication*. Toronto: Key Porter Books, 1994.

Næss, Arne. "Self realization: An ecological approach to being in the world," in *Thinking like a mountain: Towards a council of all beings*, eds. J. Seed, J. Macy, P. Fleming and A. Næss. Gabriola Island, B.C.: New Society Publishers, 1988, 19-30.

Næss, Arne. *Life's Philosophy: Reason and Feeling in a Deeper World*. Athens, Georgia: University of Georgia Press, 2002.

Singer, Peter. *Animal Liberation: A New Ethics for Our Treatment of Animals*. New York: Random House, 1975.

Strike, Kenneth. and Soltis, Joanas F. *The ethics of teaching*. New York, NY: Teachers College Press, 2009.

Regan, Tom. *The case for animal rights*. Berkeley, CA: University of California Press, 1983.

Taylor, Paul W. *Respect for Nature: A Theory of Environmental Ethics*. Princeton, NJ: Princeton University Press, 1986.

Toon, Ann. and Toon, Steve. "A Conservationists' Conundrum: In Some Places, There are Too Many Elephants." *Earth Island Journal*. Knysna Elephant Park, June 1, 2015. Accessed July 10, 2019. http://knysnaelephantpark.co.za/a-conservationists-conundrum-in-some-places-there-are-too-many-elephants/

UNESCO. "Education 2030 Framework for Action," 2015, accessed July 10, 2019. https://www.sdg4education2030.org/education-2030-framework-action-unesco-2015

Warren, Karen. "The power and the promise of ecological feminism." *Environmental Ethics* 12, no. 2 (1990): 125-46. https://doi.org/10.5840/enviroethics199012221

Plumwood, Val. "Nature, Self, and Gender: Feminism, Environmental Philosophy, and the Critique of Rationalism." *Hypatia* 6, no. 1 (1991): 3-21. https://doi.org/10.1111/j.1527-2001.1991.tb00206.x

South African Government. "Elephant Debate Not a Ploy to Reintroduce Culling," *Bua News*, October 2004.

WEESA (Wildlife and Environment Society of Southern Africa). "Great Elephant Debate," Nelspruit, South Africa August 6, 2004, formerly accessed (no longer available). http://sanwild.org/noticeboard/news2005

Weston, Anthony. A 21st Century Ethical Toolbox. New York: Oxford University Press, 2001.

Whitehouse, Anna. and Irwin, Pat. A *field guide to the Addo elephants*. Grahamstown, SA: IFAW & Rhodes University, 2002.

# CHAPTER 5

## CAN ETHICS BE JOYFUL?

**Pedagogical intent:** This chapter shifts the focus away from ethics traditionally conceived as grounded in principles or propositions that demand compliance, or duties, and can lead to actions. These more traditional positions were developed in the previous chapter. Here the principal theorist, Arne Næss, encourages us to seek bases for ethics that arise from our relationships in the world. Such relationships can inspire actions grounded in positive expressions of our experiences, and even joy. This chapter emphasises positive motivations as an important force for ethics activity in the everyday.

**Activities:**        5.1: Birds and brands quiz

                5.2: Journal making

                5.3: Clay painting

**Main theoretical resources:** Arne Næss, his experiences and philosophy, are often summed up in the term Deep Ecology. Here, however, there is a focus on the radical shift in thinking that occurred as Næss began seeking something other than duty based, or deontological, ethics as a basis for daily decisions and ethical choices.

## Where are we living? How are we living?

How we live our lives matters. Of course that may sound cliché at first, but if we dig a little more deeply into our own realities it may take on ever more significance. In some ways this exploration is a continuation of Chapter 3. In that chapter, one of the major themes was to look at cultural artefacts – like advertisements, newspaper stories, and curricula – to identify some of the values that are embedded deeply within our own cultural contexts. Sometimes those values are embedded so deeply that they become invisible in everyday life.

   In the activities presented in Chapter 3 we looked out into the world of our own experiences. The question here in Chapter 5 is what values are embedded in us? Who have we become as a result of how we have lived? Because of the things we notice? Because of what we do? Many of these things wash over us as we are marinated[1] in the culture that envelops us, but what is that culture? What relationships does it encourage? Philosophically speaking, this awareness about who we are in the world can be framed as our ontology. Another question to contemplate throughout this chapter

---

[1]    Marie Battiste, "You can't be the global doctor," 124.

is what happens when we think less about ethical duties and more about how we are living in the world? Then, can intimate relationships in the world bring excitement and joy as a basis for our practices? We will return to these questions in the discussion at the end of the chapter.

The activity presented here begins with an examination of relationships that we might have with the more-than-human world. Significant relationships are specific and particular. To reflect this need for particularity, the activity begins with a focus on birds. It could be something else. Here an educator might better choose something that they have an interest in, or maybe something that is especially important in their own place – flowers, mammals, reptiles, snowflakes, sand, or geological features. In a sense, the activity explores places, people, and more-than-human living and nonliving things that we have begun to establish relationships with – often unwittingly.

## LEARNING ACTIVITY 5.1: BIRDS AND BRANDS QUIZ

1.  A sense of gravity is added to this activity by setting it up like a pop quiz. There are many ways to do this, but a common one would be to distribute a small piece of paper and ask students to number lines on this paper from 1 to 20. This is generally enough to create a sense of heightened attention.

    In preparation for this learning activity, you will need to collect some examples of corporate branding. That is, the logos and symbols that are used to catch people's attention and build relationships. These can be gathered up using internet sources, magazines and newspapers, or photographs from billboards and other cultural artefacts. As an example, one particularly ironic photograph was taken on the Via Di Propaganda in Rome (See Figure 5.1).

Figure 5.1  Spotted on the Via Di Propaganda
(Photo by author, Bob Jickling)

In this case the corporate name "McDonalds" would be covered or cropped out to just leave a visual image of the corporate logo – the golden arches. Perhaps more easily, corporate logos can be found on the internet for companies such as Pepsi (See Figure 5.2), Shell Oil, Mercedes and many more.

Figure 5.2  Pepsi logo (Photo by author, Bob Jickling)

Sometimes, to add a little twist, it can be good to include logos of popular sports teams such as the New York Yankees, Real Madrid, or FC Barcelona. We often like to include the logo of at least one environmental organization, for example here we use the World Wide Fund for Nature (See Figure 5.3).

Figure 5.3  World Wildlife Fund, World Wide Fund for Nature
(Photo by author, Bob Jickling)

In the end, assemble 10 images of corporate logos that would be familiar to people in your region.

Next gather images of 10 common birds from your region. Again the internet is a good source. We say common birds because the point here isn't to be deliberately difficult. Choose birds that can typically be found near the place of learning where you work, volunteer, or parent. Some judgment will be required to make an interesting selection. Now you are ready for the quiz.

First present the 10 images of the corporate logos, then the 10 images of common local birds asking the group to name each one. Then go over the answers. Of course this is not a real quiz, but rather an activity designed to open a window on our life's relationships. Scoring the mock quiz is not as important as helping students to see what knowledge they have picked up in daily journeys through life. We typically find that the corporate marketplace is much more familiar than what is going on in the fields and forests nearby. This response may not always be the case but, in an increasingly globalised and electronically connected world, it appears to be a trend.

John Willinsky,[2] a Canadian educator and activist, once said in a lecture, "literacy happens at the busy intersection of commerce and state". This analysis is often revealed in this activity. There is a kind of literacy that is accrued through life's experience. Left to take its own course, it appears that one could say that commercial interests and neo-liberal policies of the many governments that prop up these interests have a disproportionate amount of control.

Following this exercise, many themes are opened for discussion. Why do the results turn out this way – that is, a generally better recognition of corporate logos than of local birds)? How do you feel about that? What would you propose doing?

Ideally, individual and group projects might arise through discussion of the questions above. However, some extension activities, such as those posed below, may also be helpful.

## EXTENSION ACTIVITY

2. Try field trips – go outside. If people are alienated from the more-than-human world, if this world is more-or-less alien to their literacy, then take them to it. Just being there is a great start. Field trip sites can be as close as a schoolyard or an urban park. They can be to hybrid spaces where the wild and the natural meet; or they can be sites of some controversy.

   It is worth noting that these experiences can be inherently positive, as a graduate student once asked, "Why are students so happy when you take them outside somewhere for a field trip?"[3] Perhaps these

---

2   This was during a public lecture at Simon Fraser University around 1990.
3   Bob Jickling, "Sitting on an old grey stone," 172.

kinds of experiences – well conceived and conducted – can provide at least one basis for the kind of joy that Arne Næss talks about. The idea that joy might form a basis of an environmental ethic will be discussed at the end of this chapter.

Remember, many people, even from a young age, are imbued with trappings of a material culture. It takes time to engage the side of their experience that has not been more fully nourished. So, go outside often and spend time preparing students about how to be comfortable outside in the conditions you are likely to face; outdoor learning may be compromised if participants are unable to engage because their attention is focused on their own comfort (too cold, too hot, many bugs, etc.). Sometimes outdoor activities can be used to support concepts taught inside. This can be a starting place. A more interesting challenge is to find ways to listen to the more-than-human world, to begin developing a stronger relationship with a place, to let nature be our teacher, for its own sake. A goal with field trips could be to exclusively do things that cannot be done in a classroom.

One idea for a simple series of field trips is to have each learner identify a small space that will become his or her ongoing special place, easily accessible and near their place of learning. Sometimes these are called "sit spots". At intervals throughout the yearly cycle, time can be allotted for them to return to this place. Meditations, thoughts, reflections, observations, poems, drawings, and/or paintings can be produced and gathered at these times. A further option could be to collect these gatherings into a yearly almanac. What other activities could be added to this format?

A variation on the last theme would be to take the class to a manageably close field trip site, say once a month for a year. Again, a variety of activities could be developed, but at heart the development of place-based relationships requires particularity, intimacy, and repetition.

3. Thematic trips and going further afield. Another approach to field trips is to be thematic. Picking up on birds again, a way of developing sustained relationships with the more-than-human world is to approach a theme from different angles. On-site bird feeders[4] and birdbaths may be the easiest field experience to organise. An extension of this would be to explore nearby birding sites. These sites generally center on good habitat and paying attention to birds' homes is another dimension of becoming more present in the natural world. What then makes a good habitat? One associated activity can be a class or group project to develop a shared field guide to the most common birds in an area. Each member of the group could be assigned one or two birds and a compilation of entries could be placed in a binder and taken on

---

[4]  It will be important to check local protocols for bird feeding to ensure that the food put out is appropriate and that attracted birds are not endangered by predators.

field trips.

A more ambitious field trip would be a visit to a bird research site. Research is often tied to migration patterns and can sometimes involve bird banding. When a good relationship is developed with the researchers, opportunities for up-close observations of birds can occur, especially as the birds are released after banding.

There are at least three key things to remember here, and one caution. First, while the goal may be to become familiar with – even to be intimate with – the local birds, the experience of being with the bird comes first. Arising from that experience will be the names of the species, stories about the birds – personal, traditional, and scientific – and an intimate understanding of their homes and habitats. The goal here is to find some joy in the learning and not just the drudgery of memorising a list of names.

Second, these exercises are also about being present in a place. Studying birds is an entrée into habits and practices that reflect another way of being in the world.

Third, birds are not the only possible subjects. They have been used for illustrative purposes. You might prefer to focus on flowers, geology, or traditional practices in a particular landscape.

Finally, urban spaces can be excellent places for environmental education. Bird feeders, local parks, and bits of undeveloped forest and field can be enormously interesting, and filled with abundant wildlife. They can be places to nurture intimacy with the more-than-human world. However, use these spaces critically; they are already colonized by humans and domesticated animals.[5] For example, what species are present? Which species are indigenous and which are feral? And, of course, what species are absent because their habitats are destroyed, or because they have been extirpated for some other reason. It is also interesting to note that plants in some urban parks are chosen for their resistance to dog urine. What does that tell us?

4. Do not ignore social issues. The object of this section is to begin to feel what it is like to travel through the world in a different, and more intimate way. That includes our relationships with people. Field experiences can, for example, involve various kinds of service-learning projects. We can learn about social issues in a scholarly way – about their causes and consequences, removed from the lives of those involved, but this seems a very partial kind of knowledge. Working directly with NGOs and other service providers can create a completely different kind of understanding.

Importantly, social issues and environmental issues are entwined in complicated relationships. Sometimes, it can be difficult to think about environmental issues when basic human needs are threatened.

---

[5] Michael Derby, Laura Piersol and Sean Blenkinsop, "Refusing to settle for pigeons and parks," 378-389.

That does not mean that people do not care about the natural world, but they do need to begin by working on their most basic priorities. Community-based projects can reveal the complexities of social-ecological relationships in an intimate way.

5. Spend time with Elders. Elders often have profound relationships with social and environmental dimensions of the community they live in. Their access to these relationships is often based on personal experience and oral history – that is, their lives lived. Visiting Elders and helping them with daily tasks can be mutually important. However, when planning such experiences find out what standards of respect are required and what protocols should be observed.

## Putting Attentiveness and Imagination into Place

As authors, field experiences have always been part of our pedagogical practices. Sometimes these experiences focus on natural settings, sometimes in urban locations and others farther afield. Some involve physical activity like walking or canoeing. Our field experiences have also engaged in a variety of social issues through service-learning experiences. However, at times, we have also wondered if these field experiences were too filled with activity such that our groups passed through places without sufficient attentiveness. Did we remain tethered to our own socially created bubble? Did we travel too loudly and clumsily without etiquette?

Put another way, when released from the confines of structured indoor spaces learners become, understandably, social and often joyful about this liberation. Attention can be diffused and the learning opportunities can be overtaken by richness of the real-world environment and the sociability of the group. To be fair, this is important too. However, without attention the atmosphere can become more like a picnic than a field-learning experience.

So, we began a series of experiments. We wanted our groups to have more embodied experiences in the landscapes and social settings of their journeys, to be more attentive, to be more sensuously engaged. We wanted them to be more present. And we wanted to find ways to express these experiences. In the activities that follow we briefly introduce journal making, place-based art, and pinhole photography as experiments that we have tried. But in the spirit of experimentation, we encourage readers to seek, develop, and try other approaches.

## LEARNING ACTIVITY 5.2: JOURNAL MAKING

1. We like to encourage a creative and hopefully non-dogmatic approach to journal making. The aim is to avoid the drudgery that some experience

in keeping a daily diary. Rather, we prefer to employ our journals at irregular intervals, special times, or teachable moments. Or we leave a little room for our students to spontaneously engage with their journals when a special thought or experience catches their imagination.

To enhance creativity, encourage mixed forms of expression. On the writing side, try including prose, poetry, aphorisms, free form fragments, and other creative approaches. However, also include visual representations. Try drawing, watercolour painting, collage, or even Matisse-like cut outs.[6] The sky's the limit. Experiment and have fun.

A simple way to bring this together is to work on complimentary pages. One page is written while the opposite page is a visual representation. Sometimes the words will come first; other times it will be a visual image first. Another tip for encouraging learners to "bond" with their journal is to use it for everyday note keeping as well as for documenting encounters with the more-than-human world. We sometimes encourage students to work front to back for their nature reflections and start at the back page to jot notes for classes, appointments, recording contact information, etc. Saving these everyday notes can encourage people to keep their journal with them more often than just for "nature journal time". It may also germinate interesting and unlikely connections between nature reflections and the jottings of everyday life that are most often considered separate from the more-than-human world.

Throughout all of this is the importance of slowing down, more fully engaging the senses, and being evermore present in a place. A drawing in a journal can be worth a hundred photographs because of its potential to reflect the total experience of more fully being there.

On a practical note, try to use journals with no lined paper. This invites visual representations. Many small journals or sketchbooks are available. Journals can also be made from more readily available sheets of paper and simply stapled or sewn together. Another approach might be to make a series of postcards from cut up sheets of paper – text on one side and drawing or painting on the other.

## EXTENSION ACTIVITY

2. Pinhole photography. Another exciting kind of representation can be made using a pinhole camera. This is typically a homemade device.

---

[6]   Henri Matisse was a French artist who, later in life, began creating a kind of collage work. He first painted scraps of paper in vivid colours. These bits of paper were then cut into simple shapes that abstractly assembled in the essential shapes of his subjects. These shapes were then assembled to form lively compositions. His subject matter often came from the natural world. Examples of these cut out compositions can be found on the internet. Emulating this technique can provide a popular activity for all ages.

There are no viewfinders or light meters – just a simple camera with a small hole, instead of a lens, to allow light to reach photo-sensitive paper, and a photographer's finger (or maybe a bit of black tape) for a shutter. The beauty of this process is that the photographer must be more fully present during the photographic process – sensing the light conditions and judging exposure times. He or she must learn to see without staring through a viewfinder. This, too, requires patience and enhanced presence in a place. There is no distraction, or immediate gratification, by constantly reviewing the incessantly collectable digital images of our present era. It takes a little more work and imagination to make a photograph this way.

This can be a good activity for the educator who wants to bring simple aspects of technology and/or science together with aesthetics and experiential education. There are lots of "how-to" guides on the internet, and elsewhere, for those who wish to try this out.[7] You can also contact local photography clubs to seek guidance.

3. Try some basic drawing classes. Many people, educators and students alike, are intimidated by making art. They typically claim that they "cannot draw", and are clearly self-conscious about trying. We have found, like most things, that learning in this realm comes quickly with practice. While the aims of the exercises in this chapter are not aimed primarily at artistic products – the process is critical here – improved skills can improve willingness to try and enhance the excitement that journal making can bring. Exposure to just a few basic drawing techniques can very quickly encourage confidence. The concepts of "gesture", or "contour" drawing can soon help the reluctant art maker.

Gesture drawings are done very quickly, often in less than a minute. The object is to capture the general shape of the subject or its essence without overthinking the process or getting bogged down in details. Such a minimalist drawing can be seen as a "gesture". Interestingly, we can also think of sitting down, being present in a place, and making a drawing, as offering a caring gesture to the land. A contour drawing is made by drawing the outline of a subject or landscape without lifting the pencil from the paper – and generally without looking at the paper.

Building upon basic art classes, skilled educators can easily employ new techniques in journal making activities. Or, maybe, arrange for an art teacher to do a guest lesson or two in your class. Such a small investment can be a big help.

---

[7] See for example: Chris Keeney, *Pinhole Cameras: A Do-It-Yourself Guide*; and Brian J. Krummel, *The Pinhole Camera: A practical how-to manual*.

## Place as Art Space[8]

Eco-art is popular in many educational settings and this section considers the practice of eco-art as an activity to enrich and enliven one's relationship with nature. Eco-art is also known as land, environmental, or site-specific art. A good starting definition is "art that helps improve our relationship with the natural world".[9] The elements of this type of artmaking are complex and varied, and so it can be useful to get insights from practicing artists. The following artists are presented because of the integrity within their practice and a long commitment to the work.

In the 1960s, well-known British artist Richard Long was in art school and realised he wanted to take his art making out into the wide world. A *Line Made by Walking*[10] was created in a field in Wiltshire, England, where he walked backwards and forwards until the flattened grass became visible as a line. He photographed this, recording his physical intervention on the landscape. This piece is considered revolutionary for two reasons. First, it added a three-dimensional quality to his work previously assumed inherent only to sculpture and, second, it made a strong case for the ephemeral in artwork and performance art. Long was demonstrating and promoting the idea that only a light touch on the earth's surface is necessary to make land-based art and this is important in setting a historical context for eco-art. A *Line Made by Walking* also emphasised how place can be integral to the work and creative process. Long established himself as a pioneer in what is commonly known as "land art" in Britain.

Currently, Andy Goldsworthy is the artist most associated with eco-art. His work often has an ephemeral quality, either being swept away by tides, blown away by a strong breeze, or floated away as a line of red leaves that drift down a river to break up and disappear. Goldsworthy's work has strong aesthetic appeal. Perhaps even more importantly, his work realises the importance, and inseparable part, that *place* can play in the making of eco-art. As he says,

> For me looking, touching, material, place and form are all inseparable from the resulting work. It is difficult to say where one stops and another begins. Place is found by walking, direction determined by weather and season. I take the opportunity each day offers: if it is snowing, I work in snow, at leaf-fall it will be leaves; a blown over tree becomes a source of twigs and branches.[11]

8    This section was prepared by Vivian Wood Alexander, Art Educator, Thunder Bay Art Gallery, Canada.
9    Green Museum, "What is Environmental Art?"
10   Richard Long, A *Line Made by Walking*.
11   Andy Goldsworthy, *Andy Goldsworthy: A collaboration with nature*, 161-62.

Goldsworthy's piece *Limestone Cones*[12] were cone shaped stone forms. The cone appears many times in his work and the film *Rivers and Tides*,[13] illustrates how Goldworthy works with material from the land to find its limits and possibilities. His connection and sensitivity to the material and the land are palpable. Much of his work is understated and subtle and if there is a message, he is letting the work speak on its own terms.

Artist Nils-Udo's expressed his concern about the ultimate difficulties in creating work that has a light touch. Here he describes an early conundrum with his work:

> Turning nature into art? Where is the critical dividing line between nature and art? This does not interest me. What counts for me is that my actions, Utopia-like, fuse life and art into each other. Art does not interest me. My life interests me. My reaction to events that shape my existence.[14]

Nils-Udo's long career began in 1960. Throughout, he looked hard at the ethical question of working in nature and the ultimate contradiction between intent and outcome. With what he calls his "plantings" he found an ethical ground and creative process he could live with. He planted a huge spiral of various corn species in France 1994 to celebrate the 500 year anniversary of the introduction of corn from the Americas. More recently in Moncton, New Brunswick, Canada, *Entrance into Nature* was created with rocks and planted bushes. Here, Nils-Udo seems to be making an offering of nature back to nature. Although some of his work is permanent, much has a life cycle and may break down or grow up and then die. The idea of having a "light touch", or avoiding permanent or lasting destruction is often an important element in eco-art.

Lynne Hull took her work in a slightly different direction and coined the term eco-atonement. As Lucy Lippard observes, "She puts her sculpture at the service of wildlife, predicting their needs (with the expert help of biologists and other scientists) and projecting aid into sculptural form."[15] For example, Hull's sculptural work for riparian nesting habitat was created with sculpted tree-like forms, and used by hawks for years. She has also carved shapes into rock surfaces to create water holes as a way to capture rainwater for wildlife. For these art projects, Hull worked at the intriguing and exciting intersection where science and art work together.

Basia Irland is a sculptor and installation artist, a poet and book artist, and an activist for water issues. Her series *Ice Books: Receding/ Reseeding*[16] takes river water and freezes it in the shape of a book. She places "text" in the book by embedding seeds in the ice. The book form,

---

12  Andy Goldsworthy. "Limestone Cones 2005-07."
13  Thomas Reidelsheimer, *Andy Goldsworthy rivers and tides.*
14  Nils-Udo cited in John Grande, "Nature works," 59.
15  Lucy Lippard, "Resume for Lynne Hull."
16  Basia Irland, *Ice Books: Receding/Reseeding.*

in ice, encompasses the ephemeral by melting, but the seeds represent the opposite, the hope being that the seeds take hold and grow. Irland has worked to draw attention to the erosion and pollution of watersheds and her work embraces another form of eco-atonement in its restorative function. As a popular, successful eco-art project, her venture has been carried out in different places around the world and in collaboration with groups as a community art project.

Peter Von Teisenhausen is one of the best-known eco-artists in Canada. His work often has an ephemeral quality and it exemplifies the importance of place. Von Tiesenhausen's respectful relationship with nature permeates his work. People are also part of Von Tiesenhausen's understanding and practice; they collaborate as viewers and participants. His work, *Passages*,[17] consisted of 100 hand-sized boats carved, charred, painted, and filled with organic matter collected from the banks of the Bow River. These boats were then dropped into the river by one hundred members of the community in June 2010 in a ritual ceremony as part of the *Celebration of the Bow River* 2010 event.[18] This kind of community art project also draws together elements of eco-art as environmental action – citizens and visitors of Calgary could celebrate and reflect on the precious nature of the Bow River, and of life and water cycles and life forms supported by it. Von Tiesenhausen's inventive work probes the essential relationship between humans and nature.

Mel Chin planted hyper-accumulator plants in soils of a dumpsite in St. Paul, Minnesota, for the art piece *Revival Field*[19] (1991). These accumulator plants were to absorb toxins in the soil to clean it. This is an example of restorative art that pushes beyond art as object making to art as ecological activism, and collaboration with science. Chin worked with a scientist to work out the concept for this piece, yet the plans for *Revival Field* are now kept in the Public Art Gallery in Minneapolis illustrating its acceptance as a form of art.

The work of artists described here illustrates the broad scope and possibility within this genre of art making. It can fuel the imagination, encourage exploration and development of art projects wherever you live. Through these projects it is also possible to see a connection between art making and environmental ethics in action.

Taken together, it is clear that eco-art is often concerned with a different aesthetic. Much of the aesthetic in eco-art is inherent in the process and in the eyes of some, the process of producing ephemeral work that ultimately is reabsorbed back into the land is in itself beautiful. With this reframing and expansion of our understanding of aesthetic qualities,

---

[17] Peter Von Teisenhausen, *Passages*.
[18] City of Calgary, *Celebration of the Bow River 2010 Art Catalogue*.
[19] See Mel Chin, *Revival Field*.

the examples above open up myriad possibilities when we consider that eco-art:

- Is embedded in the particularity and intimacy of place; place-based activities are potent agents for being in the world differently and developing new kinds of relationships within places.
- Can be enriched through opportunities for solitude and reflection in natural – and other – places; these activities open possibilities for developing respect, wonder and care, and relationship building.
- Is a vehicle for cultivating adventure, exploration, and experimentation; and of course, the learning that arises when these activities are encouraged.
- Opens possibilities for new aesthetic considerations; art can be more than, or different from, aesthetic products; eco-art celebrates beautiful actions, too.
- Embraces social-ecological activities and beautiful acts of restoration and atonement. Eco-art can be a positive, proactive, and joyful response to ethical being.

The following activities suggest a few starting points.

## LEARNING ACTIVITY 5.3: CLAY PAINTING

1. This activity is very versatile and can be done almost anywhere with these basic requirements: an outdoor space, and for each person, a small pebble sized piece of clay, a brush with fairly stiff bristles, a small container to mix the clay paint in and a small amount of water. Clay is a universal substance and is surprisingly easy to find in disturbed areas such as ditches, riverbanks, or anywhere under construction, as well as in stream or lake beds.

   The activity will involve using a paintbrush and clay as paint so that the drawing or mark making will eventually disappear.

   The act of painting in this case is not about drawing a picture, and it is not about simply working on a different kind of canvas. Make a choice about where your marks will be made, on which surface in the outdoor space you have chosen. Finding smoother surfaces to work on is helpful; rocks, saplings (where the bark is smooth) or leaves. It is always a good idea for the facilitator to model aspects of this activity: the act of painting, the importance of looking for a darker surface to provide contrast to the clay which lightens to almost white when it is used, and getting used to and applying the clay paint.

   Begin by mixing water with clay to create paint, a thick creamy consistency is ideal. Interestingly, a significant amount of clay that is mined is used in the manufacturing of commercial paint. Some more tips on using clay:

- Clay – clay is a universal material quite easily found where the earth has been disturbed – often along stream or riverbanks, or ditches or at the site of any construction.
- Design – use simple shapes; circles, lines – no images that are recognisable as objects.
- Brush – a brush with flat stiff bristles – not a watercolour brush.

Encourage the spirit of experimentation. The experiment that resulted in this activity was to find an eco-arts practice. We have just looked at or discussed the work of several artists that work in this genre. The work is varied – there is no particular material, skill, or technique to learn. These vague boundaries encouraged the setting of four rules as a starting point. The creative work will:

- take place outside,
- use materials found at the site,
- use no or few tools, and
- must, in time, disappear and leave no mark. What rules could you make to guide your search for a different eco-arts practice? Does it matter if you break them?

The activity should be kept short and simple. Whatever the participant chooses to paint – it should not smother the object. Keep in mind the approach of adornment. You are adding to, talking to, not simply drawing on the surface. These are subtle and perhaps difficult points to introduce but it helps situate your position with nature and this is important. In some ways it echoes what John Grande describes of Nils-Udo – he "embroiders" his work onto the landscape, he makes an offering "of nature back to nature". We make our mark on the landscape, we are in the landscape – we are not drawing on paper looking from the outside.

Art is work that feels like play and play that feels like work. What is your expectation with an arts practice? How are eco-artists challenging ideas about art making?

We can see and feel through our own practices. We can also test our abilities to incorporate art making as a process that helps in our journey towards ecological sustainability and atonement. We can aim to mark our presence quietly, even secretly. Put in the time and something will appear. We cannot say, in advance, what will appear; a part of all projects is out of our control. Education is often about surprises – they enliven us. And, when relationships with nature are enlivened, surprising results can be produced.

2. Clay is found universally around the world but if it is not prominent in your location, you might want to think more broadly about mark making and look for alternate materials. If you live in a cold climate, you can try the activity above using snow as an artistic palette. In other locations it might be easier to work with sand or dirt. Experiment with

the medium that is most suitable where you live. In what ways does this activity need to be adapted to suit the new material? What new opportunities arise out of using these alternative media?

## EXTENSION ACTIVITY

3. Andy Goldsworthy is considered by some the gold standard for eco-art education.[20] The documentary film about his work, *Rivers and Tides*[21] has effectively promoted eco-art in many educational settings. It makes this style of art accessible to learners of all ages. First, as a group, watch the video *Rivers and Tides*. It is widely available; however, if the whole video cannot be located, many segments of it are available on YouTube.

    Take a short walk around an area you would be comfortable working in. See what natural materials you find as you wander and consider simple patterns and designs. Touch things, compare, decide what you would like to work with and then get to work. Keep it simple: a line, a circle, keep it non-literal – and as an educator be open to whatever emerges. If you Google eco-art by students you will easily find many examples. Take a photo to remember it. If it seems appropriate, capture words and/or images about your eco-art in your journal.

    "I take the opportunities that each day offers... I stop at a place or pick up a material because I feel that there is something to be discovered. Here is where I can learn." – Andy Goldsworthy[22]

4. Review the artists in the introduction to this section.[23] Find images of their projects on the internet. As a group exercise, imagine how you can collectively make a contribution to a community – how a project could make a difference socially and/or environmentally – and reach a consensus about what eco-art project should be undertaken. What additional ideas can you add, or can you apply to something already accomplished in your community? Will you want to involve active engagement with the broader community? How can this project become a beautiful action?

---

[20] See for example: Hilary Inwood, "Shades of green," 33-38; Hilary Inwood, "Artistic approaches to environmental education."

[21] Reidelsheimer, *Andy Goldsworthy rivers and tides: Working with time*.

[22] See for example Andy Goldsworthy, *Andy Goldsworthy: A collaboration with nature*, 161-62.

[23] If available, the book *To Life: Eco art in pursuit of a sustainable planet* by Linda Weintraub has a collection of artists from around the world involved in an even broader range of projects that will further spark ideas for eco-art projects in a community setting.

# SOME THOUGHTS ABOUT THEORY

## Putting an ecological ontology back into ethics

The theoretical grounding for this chapter begins with the work of Norwegian philosopher, Arne Næss. Most importantly, what is at play here is a radical break that he is making from those authors and practitioners who are seeking to develop environmental ethics by extending human moral frameworks. As described in Chapter 4, this moral extensionism had typically rested on compliance with concepts like codes, operating rules, fundamental principles, duties, moral obligations, and categorical imperatives. This appeal to duties as both an inspiration for action, and measure of compliance, is philosophically termed *deontological* – that is, rooted in obedience to duty.

One thing that makes Næss so interesting in the evolution of environmental ethics is that he was looking for an approach to eco-philosophy, or ethics that did not appeal to abstract principles and duties. He was dubious about people's willingness to sacrifice their own interests out of a sense of duty to a larger purpose, or in order to show love of nature. For him, guilt from falling short of such lofty goals is a poor motivator.

Næss drew upon philosopher Immanuel Kant's distinction between *moral* acts and *beautiful* acts to frame alternatives. Put this way, moral acts are performed out of respect for, and obedience to, moral laws and duties. As Næss says, "the supreme indication of our success in performing a pure, moral act is that we do it completely against our inclination, that we hate to do it, but are compelled by our respect for moral law."[24] On the other hand, if we do something that is right – that is with respect to a moral law – but because of a positive inclination, then we perform a beautiful act.

For Næss, environmental ethics shifts away from moral acts and deontological ethics towards finding ways to enable people's inclinations, to enable beautiful acts. He was looking for a way of discovering the world that was positive, where people would act out of joy, fondness, and empathy rather than duty or guilt.

Of course Næss was not the only philosopher concerned with moral codes. Although they may not have been cut from the same argumentative cloth, his ideas resonate with those of John Livingston[25] and Zygmunt Bauman.[26] For Livingston, moral codes are unknown in nature and, as human creations, are more like prosthetic devices. For him it is important to develop an extended consciousness beyond the mere self. Bauman argues that complying with moral codes reduces responsibilities for one's own moral actions; codes actively erode our moral impulses. For Næss and

---

24    Arne Næss, "Self realization: An ecological approach," 28.
25    John Livingston, *Rogue primate*.
26    Zygmunt Bauman, *Postmodern ethics*. (Oxford: Blackwell Publishers, 1993); Zygmunt Bauman, *Does ethics have a chance in a world of consumers*.

Bauman, it is more important to find ways to develop people's inclinations, or moral impulses, than to develop their morals.

Næss's encouragement to find joyful ethics is perhaps his biggest gift and his greatest challenge. His own methodology for exploring this task was through a process that he called Deep Ecology. In its original sense, Næss wanted people to ask deeper, more philosophical questions as a means of exploring their most fundamental assumptions about their relationships within their own ecologies. In this sense, he is using ecology in its original sense as *oikos*, or household, or our place in the world. Once revealed, these assumptions could be examined and acted upon. This process is something that Næss encouraged everyone to do for themselves.

As a process, or methodology, Deep Ecology can be developed into another exercise in the form of evermore probing questions. This process might begin with a question about a particular action in the world. For example, why do you care for a particular thing, place, or relationship? Through persistent questions like "why?" or "for what reasons?", participants can be encouraged to probe more deeply the reasons and intuitions that underlie these choices. Eventually, this process of questioning can lead a participant to a point where she or he cannot reach any more deeply – they have reached a bedrock of sorts. The answers will vary but will typically reveal an individual's ultimate philosophical or religious idea, or perhaps their deepest intuitions.

For Næss, the Deep Ecological process – combined with rich life experiences – led him to realise that his self, or his being, was an interconnected part of the wondrous ecology – or *oikos* – in which he lived. For him this was joyous, and "part of the joy", he has said, "stems from the consciousness of our intimate relation to something bigger than our own ego, something that has endured for millions of years. The requisite care flows naturally if the self is widened and deepened so that protection of free nature is felt and conceived of as protection of our very selves."[27] By placing himself in the world in a way that opened himself to the possibility of these insights, he was also engaging in a particular way of being in the world – that is, a particular ontology.

For Næss, his personal philosophy – that he calls Ecosophy T – involves the expanding of the small ego-centric self into an expansive Self. Further, this large "S" self is in relationship with the world around him or herself; this involves being in the world in ways that enable awareness and intimacy. Philosophically speaking, thinking about being in the world is termed ontology. In this case, Næss's ontology (toward the expansive Self) is different from that of Cartesian and scientific legacies that tend to constrain *being* internally – an ontology that separates minds from bodies and a world outside. This is the basis for standing aloof, disinterested, and objective. For Næss, an ecological ontology entails engagement

---

[27]   Næss, "Self realization: An ecological approach to being in the world," 29.

with the world in an intimate and joyous way, and in a way in which we see ourselves as more than minds encased in skin and bones. In this ontological positioning, the Self is expanded to include one's relationships in the world, which are importantly mediated through the body.

In his own words Arne Næss describes the identification process as follows:

> What would be a paradigm situation of identification? It is a situation in which identification elicits intense empathy. My standard example has to do with a non-human being I met 40 years ago. I looked through an old-fashioned microscope at the dramatic meeting of two drops of different chemicals. A flea jumped from a lemming that was strolling along the table and landed in the middle of the acid chemicals. To save it was impossible. It took many minutes for the flea to die. Its movements were dreadfully expressive. What I felt was, naturally, a painful compassion and empathy. But the empathy was *not* basic. What *was* basic was the process of identification, that "I see myself in the flea." If I was alienated from the flea, not seeing intuitively anything resembling myself, the death struggle would have left me indifferent. So there must be identification in order for there to be compassion and, among humans, solidarity.[28]

Næss repeatedly points to this experience as one that has shaped the contours of his thinking, and perhaps we can even say his transformation. In this recognition of suffering, he began to see the world differently. He recognised the flea's suffering and his own identification with it; this was the basis from which he went forward for more than four decades.

What Næss has done is a radical reimagining of ethical possibilities. He is suggesting that environmental movements will be more likely invigorated by an ecological ontology and a higher realism than by an ethics based in abstract, deontological duties and principles. He believes that we will change our behaviours more readily through encouragement and a deepened perception of reality and our relationships with our places. Finally he adds, "We need the immense variety of sources of joy opened through increased sensitivity toward the richness and diversity of life, through the profound cherishing of free natural landscapes."[29] We hope that the journaling and eco-art activities in this chapter can open up some of these sources for joy and sensitivity.

## Understanding Næss in context

At this point it is worth taking a brief look at the Norwegian context in which Næss was working to see what effect the social and political context may have had on this work. Most Norwegians grow up imbued in a culture

---

28  Ibid., 22.
29  Ibid., 29.

of *friluftsliv*.[30] Literally translated, this term means "free air life". Though, it is one of those terms that, in its own language, is imbued with values and respect for the landscape. As Næss says, it is a kind of outdoor activity "that seeks to come to nature on its own terms".[31] In Norway it is closer to a way of life, a way of being that Næss speaks about.

Of course, powerful forces encroach on this traditional concept of *friluftsliv*. There is much discussion about co-option by consumer forces, competition, and other environmentally destructive intrusions into nature. Many point to Norway's oil industry as a force inconsistent with *friluftsliv*. Still there are several attributes that can be taken from this cultural phenomenon. First, weakened as it may be, it still stands a modern version of an alternative lifestyle. As people seek new ways of living and being in their own landscapes, it may be useful to look at what the Norwegians have been doing.

Second, as environmental discourse shifted emphasis to the social influences in the construction of nature, many Norwegians and their followers insisted on a realist element in their environmental thought. This seems to have enabled the development and persistence of Deep Ecology and Næss's concept of Self, and Self-realisation. This has led to a philosophical re-emergence of interest in Nordic ecophilosophy. For example, Roy Bhaskar, prominent scholar and a leader in the critical realist movement, has declared that "Nordic ecophilosophy has been generally much stronger, earlier, deeper and more persistent on ecological matters"[32] – more than critical realism. Still, these critical realist philosophers are poised to advance these early ecophilosophical developments. Some of these possibilities are taken up in Chapter 10.

### Similar critiques of environmental ethics

Feminist scholars have also troubled the notion of a universal code of ethics. Classic works by writers such as Karen Warren[33] and Val Plumwood[34] argue that environmental ethics is not an objective or disinterested theory, but rather emerges as theory in process. According to these accounts, and other landmark writing such as Jim Cheney's[35] writing on postmodern environmental ethics, ethics needs to be contextualised, narrated, and to hold a place for feelings and emotional understanding. A rich and growing

---

[30] In some contemporary polls as many as 90 % of Norwegians self-identify as practitioners of *friluftsliv*.

[31] Arne Næss, *Ecology, Community and Lifestyle*, 178.

[32] Roy Bhaskar, "Critical realism in resonance with Nordic ecophilosophy," 9.

[33] Karen Warren, "The power and promise of ecological feminism," 125–146.

[34] Val Plumwood, "Nature, self, and gender," 3–27; Val Plumwood, *Feminism and the mastery of nature*; Val Plumwood, *Environmental culture*.

[35] Jim Cheney, "Postmodern environmental ethics: Ethics as bioregional narrative," 117–134.

body of feminist literature promises to lead researchers into interesting new territory. Some of these possibilities are taken up in Chapter 6.

## Critique of Deep Ecology

The aspects of Arne Næss's philosophy that are helpful in this chapter are often found under the broader umbrella of Deep Ecology. At the time Deep Ecology was rising in international prominence, notably with the publication of *Deep Ecology: Living as if Life Mattered* by Bill Devall and George Sessions,[36] environmental education was taking a socially critical turn. The Dark Green environmentalism often associated with movements like Gaianism and Deep Ecology was, in many quarters, being dismissed as naïve. For people like John Fien,[37] this darker shade of environmentalism was really only feasible for the world's affluent few who could afford to disentangle themselves from the economic imperatives of daily survival. For him, a red-green environmentalism was a more "realistic" option and the newly promoted education for sustainable development was an attractive platform.

In this context, it is not surprising to see concerns about Deep Ecology arising from those in the developing world. For example, Indian scholar Ramanchandra Guha wrote an early critique;[38] amongst his critical points two emerge as interesting for discussion in this context. First, he argues that Deep Ecology is axiomatic – that is it is based on basic, or self-evident truths. In particular, he is concerned about the assumed truth arising from the anthropocentric-biocentric distinction, and the associated imperative to move towards a biocentric worldview. He views the resulting emphasis on preservation of biotic diversity would not adequately account for: the disproportionate share of resources consumed by industrial countries as a whole and the urban elite within the "Third World" – the burdens that this agenda would place on the poor, the landless, and women; as well as other questions related to equity and political and economic equity.

Second, Guha asserts that Deep Ecology is uniquely American and largely focused on wilderness preservation. He then rightly critiques the imperialist yearnings of a Western preservationist culture and the social uprooting of human populations around the world.

Guha understandably and correctly criticises imperialist tendencies and consequences of colonisation in an evermore-globalising world. The diffusion of environmental thinking arising from affluent nations can feel like – and indeed be enacted as – yet more colonisation. What is less fortunate is a kind of essentialising of Deep Ecology in critiques like this. While some members of the Deep Ecology movement are clearly at

---

[36] Bill Devall and George Sessions, *Deep Ecology: Living as if Nature Mattered.*
[37] John Fien, *Education for the environment*, 45.
[38] Ramanchandra Guha, "Radical American environmentalism and wilderness preservation," 71-83.

fault, it does seem a mistake to link this movement so boldly to such a deeply flawed vision of wilderness preservation. At the heart of Næss's conceptualisation of Deep Ecology are many ideas that still seem to have the capacity to do good work.

It is also debatable that Deep Ecology is so singularly axiomatic as Guha claims. While there is much more to Næss's thinking than we have raised in this chapter, the key idea that we have built upon here is that he was looking for something other than a duty-based environmental ethic. In his own conception of Deep Ecology – and he always stressed that the rest of us had the obligation to come up with our own conceptions of this idea – inspiration and passion to act arise from experiences in the real world. He calls this Self-realisation and is a kind of identification with his more-than-human surroundings rooted in feelings like empathy. This seems to offer a potential break from the kind of axiomatic dogma inferred by Guhra.

From a feminist perspective, the idea of Self-realisation has limits and is subject to critique. In a remarkably interesting paper, Val Plumwood suggests that this idea can be a slippery one.[39] She is particularly critical of those Deep Ecologists, such as Warwick Fox, who strive to expand this sense of Self as widely as possible. For Plumwood, this aim is inclusive of identification with the entire cosmos. Plumood's key points include, first, that while recognising human continuity with the natural world is important, its distinctness and independence from us also needs to be recognised. She suggests that identifying with another to the point that we might see ourselves as indistinguishable from that other being makes it impossible to preserve a real sense of this other's well-being as opposed to her own. The ability to understand and care requires some distance.

Second, Plumwood sees expanding the scope of Self-identification comes with an increasing detachment with experienced relationships. As such, this identification inevitably becomes abstract. For her, feminist theory suggests that we must move forward in a way that can allow for both continuity and difference. In this way, she argues, ties to nature can be expressive of rich caring relationships of kinship and friendship. This once again brings us back to something fundamentally close to Arne Næss's own vision of Deep Ecology.

It is true that Deep Ecology and, more broadly, Nordic ecophilosophy, have arisen in a particular physical and cultural landscape. While we might quibble about some of the critiques, they do suggest that new ways of being and new ethical conversations may well need to arise in their particular contexts. There may be some common ideas and questions that arise from the literature on Deep Ecology that can be nurtured in other places; we expect that there is still much good work to be done under this umbrella. However, the practices of Deep Ecology, and ethics more broadly, will also need to arise in social and physical contexts. Or as Jim Cheney suggests,

---

[39]   Val Plumwood, "Nature, self, and gender," 3-27.

environmental ethics may necessarily be a series of bioregional narratives, differentiated to meet the needs of the places they originate, but threaded together by a common human capacity to live respectfully together on the earth; and this appears consistent with Næss's conception of Deep Ecology. This thought will be taken up in succeeding chapters.

## References

Battiste, Marie. "You can't be the global doctor if you're the colonial disease," in *Teaching as activism: Equity meets environmentalism*, ed. Peggy Tripp and Linda Muzzin. Montreal, PQ: McGill-Queen's University Press, 2005, 121-133.

Bauman, Zygmunt. *Does ethics have a chance in a world of consumers.* Cambridge, Massachusetts: Harvard University Press, 2008. https://doi .org/10.4159/9780674033948

Bauman, Zygmunt. *Postmodern ethics.* Oxford: Blackwell Publishers, 1993.

Bhaskar, Roy. "Critical realism in resonance with Nordic ecophilosopgy," in *Ecophilosophy in a world of crisis: Critical realism and the Nordic contributions*, eds. Roy Baskar, Karl George Høyer, & Petter Næss. London: Routledge, 2010, 9-24. https://doi.org/10.4324/9780203698846

Cheney, Jim. "Postmodern environmental ethics: Ethics as bioregional narrative." *Environmental Ethics* 11, no. 2 (1989): 117-134. https://doi.org /10.5840/enviroethics198911231

Chin, Mel. *Revival Field.* 1991-ongoing, accessed August 26, 2018, http:// melchin.org/oeuvre/revival-field

City of Calgary. *Celebration of the Bow River 2010 Art Catalogue.* 2010, accessed August 26, 2018, http://www.calgary.ca/CSPS/Recreation /Pages/Public-Art/Celebration-of-the-bow-catalogue.aspx

Derby, Michael, Piersol, Laura and Blenkinsop, Sean. "Refusing to settle for pigeons and parks: Urban environmental education in the age of neoliberalism." *Environmental Education Research* 21, no 3 (2015): 378-389. https://doi.org/10.1080/13504622.2014.994166

Devall, Bill and Sessions, George. *Deep Ecology: Living as if Nature Mattered.* Salt Lake City: Peregrine Books, 1985.

Jickling, Bob. "Sitting on an old grey stone: Meditations on emotional understanding," in *Fields of green: Restorying culture, environment, and education*, eds. Marcia McKenzie, Paul Hart, Heesoon Bai and Bob Jickling. Cresskill, NJ: Hampton Press, 2009, 163-173.

Næss, Arne. *Ecology, Community and Lifestyle: Outline of an Ecosophy*, trans. and revised by David Rothenberg. Cambridge: Cambridge University Press, 1989. https://doi.org/10.1017/CBO9780511525599

Næss, Arne. "Self realization: An ecological approach to being in the world," in *Thinking like a mountain: Towards a council of all beings*, eds. John

Seed, Joanna Macy, Pat Fleming, and Arne Næss. Gabriola Island, B.C.: New Society Publishers, 1988, 19-30.

Fien, John. *Education for the environment: Critical curriculum theorising and environmental education*. Geelong: Deakin University Press, 1993.

Guha, Ramanchandra. "Radical American environmentalism and wilderness preservation: A Third World critique." *Environmental Ethics* 11, no. 1 (1989): 71-83. https://doi.org/10.5840/enviroethics198911123

Grande, John. "Nature works." *Art & Design Profile No. 36, Special Issue: Art and the Natural Environment*, 1994: 59.

Green Museum. "What is Environmental Art?" accessed August 26, 2018, https://web.archive.org/web/20170606113917/http://greenmuseum .org/what_is_ea.php

Goldsworthy, Andy. *Andy Goldsworthy: A collaboration with nature*. New York: Harry N. Abrams, Inc., 1990.

Goldsworthy, Andy. "Limestone Cones 2005-07," Andy Goldsworthy Digital Catalogue, accessed August 26, 2018, https://www.goldsworthy.cc.gla .ac.uk/image/?id=ag_03314

Inwood, Hilary. "Shades of green: Growing environmentalism through art education." *Art Education* 63, no. 6 (2010): 33-38. https://doi.org/10 .1080/00043125.2010.11519100

Inwood, Hilary. "Artistic approaches to environmental education: Developing eco-art education in elementary classrooms." PhD thesis, Concordia University, Canada, 2009.

Irland, Basia. *Ice Books: Receding/Reseeding*. Accessed August 26, 2018. http://www.basiairland.com/projects/ice%20books/index.html

Keeney, Chris. *Pinhole Cameras: A Do-It-Yourself Guide*. New York: Princeton Architectural Press, 2011.

Krummel, Brian J. *The Pinhole Camera: A practical how-to manual for making pinhole cameras and images*. CreateSpace Independent Pub, 2009.

Lippard, Lucy. Resume for Lynne Hull, 2000, accessed August 26, 2018, http://eco-art.org/?page_id=748

Livingston, John. *Rogue primate: An exploration of human domestication*. Boulder, CO.: Roberts Rinehart Publishers, 1994.

Long, Richard. *A Line Made by Walking*. 1967. Photograph of physical intervention. Tate/National Galleries of Scotland, accessed August 26, 2018, https://www.tate.org.uk/art/artworks/long-a-line-made-by -walking-ar00142

Nils-Udo, Entrée. University of Moncton, accessed August 26, 2018, https:// www.umoncton.ca/parc-ecologique-millenaire/en/entree

Plumwood, Val. "Nature, self, and gender: Feminism, environmental philosophy, and the critique of reason." *Hypatia* 6, no. 1 (1991): 3-27. https://doi.org/10.1111/j.1527-2001.1991.tb00206.x

Plumwood, Val. *Feminism and the mastery of nature*. New York: Routledge,1993.

Plumwood, Val. *Environmental culture: The ecological crisis of reason*. London, UK: Routledge, 2002.

Reidelsheimer, Thomas. prod. and dir. *Andy Goldsworthy rivers and tides: Working with time*. US: Mediopolis Films, 2003.

Von Teisenhausen, Peter. *Passages*. 2010, accessed August 26, 2018, https://www.tiesenhausen.net/landscape

Warren, Karen. "The power and promise of ecological feminism." *Environmental Ethics* 12, no. 2 (1990): 125-146. https://doi.org/10.5840/enviroethics199012221

Weintraub, Linda. *To Life! Eco art in pursuit of a sustainable planet*. Berkley, CA: University of California Press, 2012. https://doi.org/10.1525/9780520954236

# CHAPTER 6

## WHAT IS THE PLACE FOR (MY) HISTORY, (MY) CULTURE, (MY) FEELINGS?

This chapter illustrates how our responsiveness to ethical concerns is influenced by history, culture, social circumstances, and feelings. It explains that taking ethical action is not simply a case of obediently following established ethical codes in society or even pursuing a more "joyful" ethical engagement with the world, as proposed in the previous chapter. Instead, our values and moral impulses are very often in tension with the circumstances of our daily lives and the kind of society we live in.

**Activities:**    6.1: Martine's dilemma

6.2: How should we care for dogs?

6.3: Buying organic

6.4: World map

**Pedagogical intent:** This chapter will introduce teachers to the importance of acknowledging culture, past experiences, and feelings in ethics decision- making. The chapter will illustrate how history, culture, and feelings influence how people make sense of ethics concerns, how they express them, and also how they navigate them.

## Ethical deliberations in the lived world

How do we come to relate to the world in the ways that we do, and value some things over others? What compels us to take action when we encounter something that we consider to be "wrong" or "bad"? These are questions at the heart of this chapter, which considers how history, culture, and social circumstances influence our ethical engagement with the world. We respond to these questions by proposing that each person's life experiences, thoughts, feelings, values, and aspirations occur in a back-and-forth relationship with his or her social, cultural, and ecological contexts. Even when we think we are taking a moral stand on something "alone", based on our own, internal responses, we are, in fact, responding in and through the world around us.

These ideas are neatly captured by two influential North American environmental thinkers. Lucie Sauvé writes, "We are embodied, localized, contextually grounded beings"[1] and Anthony Weston similarly explains that our values are "deeply embedded in and coevolved with social institutions

---

[1]    Lucie Sauvé, "Being Here Together," 325.

and practices".[2] Values, he concludes, are by their very nature subject to change and open to reshaping.

Part of what makes us human is our internal struggles: we weigh up and respond to the complexity of the world we experience, always in relation to what we care about most, which, as we have noted, is deeply influenced by the human and more than human world around us. Sociologists refer to this as our "reflexive agency", that is, each one of us has the individual power to learn about, deliberate, and act on how we think we ought to live our lives in relation to the world around us.

In this book, we recognise that such processes of ethical deliberation are strongly influenced by the history of the places we inhabit, the cultural practices that give our daily lives meaning and structure, and the social circumstances that we experience. We also recognise that these processes of ethical deliberation are really a type of reflexive learning that germinates action and is potentially the bearer of social transformation at a scale much wider than the individual.

## Who decides who can fish where?

It is interesting to talk about ethical deliberations in general. However, in the real world these deliberations can be complex. For example, Martine works for a local environmental NGO in a small coastal town on the south-eastern coast of South Africa. The town is built around a river estuary and surrounded by a National Park that includes a Marine Protected Area. Martine's job is to provide educational programmes for the local communities ranging from wealthy, predominantly white residents to poor, predominantly black and mixed-race residents, as well as to tourists during holiday seasons.

A few years ago, Martine participated in a course in Environmental Education. During the course, Martine wrote about her ethical struggles as an environmental educator:

> In my day-to-day work with people and communities I'm confronted with what is right according to the "books" and what is right according to myself. In my work, I have to take action according to my work and not my own ethics. For example, if a man from a nearby community comes to fish without a permit to have food on the table, how can I let that person get a fine for fishing in a marine protected area?

What we see here is evidence of Martine wrestling with the tension between fulfilling her professional obligations (i.e., engaging with local communities around the management of a marine protected area), and her deeply held personal commitment to the well-being and dignity of the poor

---

[2]    Anthony Weston, "Before environmental ethics," 27.

fisher folk living in the area (i.e., understanding that an unemployed father living at the coast will turn to local natural resources to feed his family).

Figure 6.1    Martine's dilemma (Photo courtesy of Cindy-Lee Cloete)

As an educator, Martine needed to identify these and other tensions and consider what they meant for how she worked educationally with others. Should she act on her ethic of concern for the well-being of a fisherman and his hungry family, but thereby undermine the country's environmental legislation which aims to conserve the country's natural resources for current and future generations? Should she act on her ethic of concern for the long-term integrity of the marine protected area where she works, striving to re-establish the declining fish stocks and fragile marine and coastal ecosystems, even when this means denying access to local fishermen and, in fact, issuing fines when they do not comply? If she turns a blind eye to this one fisherman, what will she do if and when others arrive too? Will she acknowledge the right of *some* people to catch fish in the Marine Protected Area, but exclude others? What values will guide her to make those decisions? Not unlike the tensions experienced by Jean Valjean in Victor Hugo's *Les Miserables*, introduced in Chapter 2, Martine must navigate around various categorical imperatives, and the lived realities of people in dire need. A fraught position, to be sure.

One of the reasons environmental problems can be so difficult to understand and respond to ethically is because they occur in social-ecological settings where people's beliefs, values, priorities, and ways of doing things intersect with the realities of the natural world. These realities can be as varied as geographical distances, weather patterns, soil types, seasonal availability of resources such as water or food crops, reproductive rates of certain plants or animals, disease vectors and so on. Sometimes, the realities have been created by people, and the systems they devise for themselves and others. These include economic systems, political regimes, and firm cultural or religious conventions that direct the way people interact with the natural world. We refer to these as "social-ecological settings" in recognition of the ways that social systems interact with ecological systems – usually in complex and open-ended ways.

## LEARNING ACTIVITY 6.1: MARTINE'S DILEMMA

1. What ethical dilemmas did Martine face? Try to explain in terms of the social, political, and economic tensions in the Marine Protected Area.

2. Martine's response to this ethical quandary was to keep communicating with the local subsistence fishermen about coastal conservation laws and the importance of upholding them, but she never fined or reported them if she found them fishing illegally in the protected area. Many would support her decision, saying that she took the best option: keeping enough pressure to deter all but the most desperate people from illegal fishing, while maintaining respectful relations and open channels of communication between nature conservators and communities. Others, however, might criticise Martine for being a "fence-sitter" who was willing to compromise the long-term sustainability of the coastal ecosystem in favour of short-term social benefits. What would *you* do if you were in Martine's shoes? Discuss your reasons.

## EXTENSION ACTIVITY

3. Martine's decision-making was based on her sensitivity to her social-ecological, socio-economic context, and her recognition that the local fishermen's environmental ethics were strongly influenced by their life circumstances. Her response reflects this chapter's emphasis on how our ethical deliberations are invariably influenced by our histories and our social, cultural, and personal contexts. However, as opened up in Chapter 1's consideration of ethical relativism, we must recognise the importance of thinking "more widely than personal experiences only, and to think about these in relation to the common good."

    What is a "common good" in this case story? Do you think that Martine's response is adequate, or are there other things to be considered when deciding if subsistence fishing in a Marine Protected Area is "the right thing to do"? Is it possible to fish, or otherwise harvest, ethically within a conservation area?

4. Do you have similar examples where you live of a community's, school's or organisation's ethical response to an environmental concern being influenced by economic, cultural, religious, or political tensions? Research a local example if you do not know of one. How have these tensions been resolved, if at all? Were they resolved "for the common good", or something else?

### A diverse world makes for diverse ethical responses

Environmental values, traditions, norms, and perspectives on the natural world are dynamic and vary according to different social, economic, and cultural settings. This makes it difficult to find a "one-size-fits-all" ethical response to urgent environmental challenges. People from different backgrounds have different understandings of the "right" way to act, and often they all raise valid points.

Different backgrounds do not necessarily mean different cultures; it can also refer to different economic circumstances, different educational backgrounds, differing levels of access to important resources and so on.

Some years ago, Lausanne Olvitt talked to a group of environmental educators from the southern African region about the things that influence their ethical decision-making regarding the environment.[3] They described how their efforts to take responsible environmental action were made in relation to factors as varied as financial concerns, time frames, social conformity, the value attached to the subject, as well as the lack of alternatives to current practices. When asked what values or codes guide them when faced with making decisions affecting the environment, the educators mostly referred to:

- **Social concerns**: "I'm guided by society e.g., what will people think about my decision."
- **Self-interest**: "I am more concerned whether I'll end up in a safe side or not."
- **Cultural influences**: "I respect nature because it was made for me, therefore it is taboo to spoil it."
- **Religion**: "As a Christian, the concept of stewardship of the earth is the most influential when it comes to my values."
- **Past experiences**: "The prolonged drought I experienced in my childhood where I had to carry a 25-litre bucket of water on my head and 5-10 litres in hands from the spring... forced me to modify habits in handling and use of water."

Similar insights were gained through conversations with a group of South African youths identified as "at risk" due to their social circumstances of homelessness, poverty, substance abuse, or gang involvement. The youths' narrations of their environmental actions showed that their actions were influenced more by their social and economic identities and past experiences than by some abstract ethical code.

One of the youths, for example, stated that she understood the importance of energy conservation and willingly turned off the lights in the hostel if she was the last to leave a room. However, she was not permitted to do so when she vacationed at the family's rural home because her grandmother wanted all the lights turned on so that others would know that the family could afford to pay the electricity bill. This is perhaps not too surprising considering that the grandmother had lived in poverty most of her life and, under the racially segregated Apartheid government,

---

3    Lausanne Olvitt, "Working with environmental ethics and adult education," 43-45; Lausanne Olvitt, "Doing What's Right for People and Planet."

electrification of black people's homes in rural areas was uncommon. So, to reach old age in democracy and receive a government pension and a house with electricity was something of immense value to the grandmother, and no granddaughter with energy-saving tips from the city was going to change her mind!

Another youth explained that he had in the past littered intentionally in the city streets with the aim of creating, or at least sustaining, job opportunities for garbage collectors in a country with remarkably high unemployment and poverty levels. "Surely", he argued, "if nobody dropped litter in the streets and parks, there'd be no work for the cleaners, and then what will they and their families do for money?" However, he reflected that, after being part of an environmental education programme, his knowledge and understanding of environmental matters improved and he felt better equipped to make wise choices about things that affected the environment.

These and other stories of people's authentic *in situ* struggles to live well in the world give us insight into the diverse socio-cultural, economic, and political spaces in which all of us negotiate ethical actions every day. Such stories offer glimpses into the complexity, uncertainty, and creative possibility of people's lived experiences. Sharing them can help environmental educators to create bridges between abstract, philosophical ideas and real-life experiences. These ideas are summed up by Arjen Wals who writes that environmental learning processes are "rooted in the life-worlds of people and the encounters they have with one another"[4] and are thus open-ended and potentially transformative. This is a significant opening for the work of environmental ethics and education.

---

4    Arjen Wals, "Learning in a changing world and changing in a learning world," 43-45.

## LEARNING ACTIVITY 6.2: HOW SHOULD WE CARE FOR DOGS?

Many of the ethically-laden circumstances mentioned so far in this chapter are so much a part of our everyday lives that we seldom "step back" to reflect on their socio-cultural origins – or the ethical complexity of resolving them. This activity focuses on the widespread problem of homeless and hungry dogs on the streets. The aim is to explore some ethical responses to the street dog problem from a range of socio-economic and cultural perspectives. After you have had a go with this generalised example, you can apply your thinking skills to your own more local and specific examples.

1. The handout "Who says we should care for dogs?" considers the problem of homeless dogs on the streets and some people's strongly held responses to them. The speech bubbles reflect some common responses. With members of your group, try to identify what social, cultural, and historical influences might lie behind each ethical response.

2. Now discuss YOUR OWN ethical response to the problem of homeless dogs. Discuss with a partner how your response may be influenced by your background, your socio-cultural context, and your feelings.

3. Think of other possible responses to the problem of homeless dogs beyond what appears in the speech bubbles of the handout. What environmental values do the responses reflect? Where might such values originate?

# HANDOUT:
# WHO SAYS WE SHOULD CARE FOR DOGS?

I will not support animal welfare organisations and I think the most sensible solution is to euthanize these poor dogs. There is so much human suffering and need in the world that I cannot morally justify donating my money to feed disease-ridden dogs. I donate to a local charity that provides a safe haven for abused children in my city, and to an international group supporting North African refugees.

We need many more animal welfare organisations, not only to care for the immediate needs of these vulnerable and suffering animals, but also to ensure there are sterilisation programmes to reduce the problem in future. I volunteer on Saturday afternoons at the local dog shelter. I clean cages and exercise the dogs. It's so rewarding to see such small efforts making a big difference in the lives of these animals!

I am an animal lover and I don't like to see them suffer. But I am poor, unemployed and can hardly feed myself some days. So, I'm sorry, but looking after animals on the streets is really not a priority for me!

I believe we should help in extreme cases of animal neglect and abuse but, first and foremost, we must discourage the multi-million dollar pet industry. I think it is unethical to keep pets when it has that kind of global environmental impact. We cannot afford to get sentimental over needy cats and dogs when the global cost is so high.

These animals are hungry, afraid and suffering because of the kind of world that we, as people, have created. Caring for these animals on the streets should be a natural extension of the care and responsibility we feel for one another. Remember Mahatma Gandhi's words: "The greatness of a nation and its moral progress can be judged by the way its animals are treated."

## EXTENSION ACTIVITY

Did you find yourself relying on generalisations or stereotypes during your discussion, for example, "rich people feel like this about dogs, whereas poor people or people from such-and-such a culture feel differently"?

Stereotyping can indeed be a challenge in classrooms when foregrounding social, cultural, religious, economic, and historical influences in people's ethical decision-making. Stereotypes are widely-held, preconceived and fixed beliefs about particular groups of people and are often talked about in terms of race, age, culture and gender equity. Stereotypes don't just affect individuals; they can also negatively affect whole communities' ability to learn about and transform environmental problems. Stereotypes create barriers to honest conversations and they deny people the opportunity to understand the uniqueness and complexities of other people's lives.

4. Think carefully about the anonymous speakers in the handout. As you read the bubbles, did you find yourself linking them to different social stereotypes? Thinking, for example, "That's the kind of thing XXX type of person would say." How might such stereotypes affect the way you and your local community collaborate around the problem of homeless street dogs?

5. Do you have any preconceived ideas about the following groups of people, their environmental values, and the reasons for their way of interacting with the natural world?
   - Rhino poachers
   - Inner city street children
   - People who throw their trash out the car window
   - Vegetarians
   - People who sign petitions to save the rainforests
   - Big game hunters

6. What led you to hold these views of these groups of people, and do you know how accurate they are? For a more in-depth discussion, consider discussing only one or two groups.

7. What would you need to know and do if you wanted to have a respectful and productive conversation about caring for the world with one or all these groups of people?

## Re-personalising morality in the light of our histories, cultures and social circumstances

The discussions and activities so far in this chapter have illustrated how socially complex ethical deliberations can be. The emphasis so far has been on how ethical tensions arise between different people or groups of people. We have considered the case of Martine who had to navigate the tensions between her organisation, the national conservation laws, the local fishing community, and her own moral response. We have also read about how people's decision-making about environmental concerns is influenced by their families, religious convictions, financial means, cultural norms and taboos, and so on. In the last extension activity, we explored the pitfalls of social stereotyping and how problematic it can be to "put people into boxes" before we've even started to engage with them about an ethical concern.

Now, in this section, we focus on the ethical tensions that can arise *within* a single person. This section reminds us that all of us are locked in internal, deeply personal struggles about what we stand for, and what counts as "the right thing to do". Even when you meet someone who takes a strong stance on a particular environmental issue, remember that his or her outward position may appear simpler and stronger than their internal moral struggles.

Many people experience ethics as a pre-determined set of rules, a code of conduct, a standardised way of judging right from wrong. Taking ethical environmental action is often portrayed as a clear-cut debate: some aspect of our culture is "wrong" and we just have to reject it and start doing the "right" thing. Such an understanding of ethics is limited because it implies that ethics is something external to us, something impersonal, static and unresponsive. More often than not, the socio-cultural practices that nag at our conscience are the same socio-cultural practices that shape our identities and give us a sense of belonging and continuity within our communities. It is not just a straightforward case of changing our ways once we realise where the problem lies. Zygmunt Bauman, a Polish-born sociologist, urges us instead to re-personalise morality by freeing it from "the stiff armour of the artificially constructed ethical codes".[5]

Part of the process of re-personalising morality involves re-examining our ethical actions in the light of our histories, cultures, and social circumstances, as well as being open to changing or defending them in the light of new knowledge and changing priorities, be it at a global or personal level.

## Taking Responsibility: Personal dimensions

This activity directs our attention to personal, moral dimensions of taking responsible environmental action. Consider the following story and the other short examples that follow:

Nkanyiso is a young Zulu man working for an environmental NGO near Durban, South Africa. He was raised in a traditional, rural Zulu family where eating meat was recognised as a sign of success and prosperity. However, in his early twenties, Nkanyiso got part-

---

5    See Zygmunt Bauman, *Postmodern ethics*, 34.

time work with environmentalists, many of whom were vegetarian. Over many months he came to understand and support the ethical stance that we should stop, or at least reduce, eating commercially produced meat due to its impact on land degradation and water quality – not to mention animal rights. Although Nkanyiso tried to reduce the amount of meat he ate, he said it was impossible for him to refuse it when he visited his traditional family home. He was also sensitive to being mocked by his young, male Zulu friends when they went to town to buy takeaways and Nkanyiso opted to buy "meat-free". Furthermore, Nkanyiso admits that he really enjoys eating meat and that the vegetarian meals his colleagues often share with him sometimes leave him feeling unsatisfied.

Nkanyiso's story illustrates how ethical actions, no matter how small, local, and deeply personal, are always influenced by socio-cultural values, norms, and the histories from which they emerge. His story highlights the importance of personal morality that is different from merely falling in line with the ethical codes and norms of society. Nkanyiso is deeply challenged at a personal level when faced with the vegetarian debate. His work colleagues, family members and friends each contribute a different perspective on the ethical question of eating meat, but ultimately it is up to Nkanyiso to decide where he stands on the matter – and follow through with morally defensible actions.

All of us experience similar moral choices on a daily basis. Below, we share some common examples, but you will certainly be able to add more examples about the tensions between what you'd "like" to do and what your circumstances "allow". Consider:

- The young, working class parents who would love to buy organic vegetables and free-range chickens for their family but cannot afford the extra cost.
- The university professor who *can* and *does* buy organically produced food, but still worries that the embodied energy in the packaging and transport of these products undermines her efforts to reduce her carbon footprint.
- The young woman who wanted to reduce the amount of meat in her diet but recognised that she'll have to wait until she is married and has her own home before being able to enact her plans. As much as she would like to, she knows that her parents and extended family will not abandon a centuries-old diet in a couple of years just to accommodate her environmental preferences.
- The father who tries to reduce his carbon footprint but stops short of trading-in his fuel-heavy, powerful car for a smaller, fuel-efficient one. His argument is that, in the city where he lives, crime levels including car hijackings are high. A few years ago, he and his family narrowly escaped being hijacked only because he was driving

a powerful car that was able to out-manoeuvre the hijackers and accelerate to safety.

Navigating the tense spaces between what we feel we *ought* to do, what we know we realistically *can* do, and how we expect others to respond to it, provides the informed impulse of ethical deliberation!

## LEARNING ACTIVITY 6.3: BUYING ORGANIC

1. With a partner or in small groups, discuss the first two examples about buying organic food, as bulleted above. The following questions can structure your discussion:
   - The young working–class parents and the university professor all express a personal moral preference for buying organically produced food, but only the professor follows through with this moral preference. What external factors influenced their respective abilities to act on their moral impulses?
   - Do you think the university professor should feel guilt or concern for the embodied energy in the packing and food miles, when she is already taking ethical action by buying organic food? How far should she take her sense of personal moral responsibility?
   - Do the working class parents have alternatives? If you were in this situation, what would you do?
   - What information and knowledge about organic food and free-range chicken would the parents, the professor and all of us need in order to make the best possible ethical choices about buying them?
   - Check yourself and your peers! Did you find yourself slipping into stereotypical views about working class people, parents, or professors in any of your responses?

## EXTENSION ACTIVITY

2. Think about a situation in your own life where you believe you know the right thing to do regarding an environmental concern, but are held back from doing it by circumstances in your home, school, university, workplace, community, or country. Jot down the basic outline of your example in a paragraph.

   Now let's analyse your situation:
   - Think about the action that you believe is the "right thing to do." What led you to hold that view? Think about the values that underpin your desired action and where they came from. Do other people hold the same or similar views, or is it only you? How long

have you felt this way about the environmental concern? What guided or influenced your view?

- Now think about the circumstances that are hindering your ethical action in response to the environmental concern. Are the constraints temporary or long-term and seemingly insurmountable? What led to them being like that and having enough power to influence what you can or can't do? Do these constraints originate in economic conditions, or political, cultural, biophysical, or some other conditions?
- In terms of education, what kinds of things should a good educational process enable you to **know** and **do** in order to follow through on what you feel is the right thing to do regarding this environmental concern? Consider, for example, what knowledge you might need, what skills would influence what you do with that knowledge?

## Navigating across land and culture: Towards right relations for decolonisation

Throughout the book, and in the current chapter, we have shown the complexity of connections between culture and the dispositions that individuals and communities come to hold as ethics. These complexities are especially present as people of different cultural and social backgrounds begin to interact. One such example of a complex socio-cultural interaction that is deeply connected to environmental ethics is the relations between and among global Indigenous communities and others who migrate to settle on land that has been inhabited by Indigenous peoples since time immemorial. Many Indigenous nations practice spirituality that is richly and intimately nested in the land on which the people of the nation depend for their survival. Indigenous land ethics are often relational in character and this quality can conflict with the rule and duty-based ethical systems that most often characterise Settler nations.[6] Moreover, and perhaps most importantly, relationships with the land which are typically highly prioritised in Indigenous ethical approaches may not be equally respected by Settler cultures.

Land, how to inhabit it well, and how to relate between Settler and Indigenous cultures are at the heart of colonial ethical conflicts. Like all ethics, we argue, these ethical relations are in need of ongoing reconsideration. Leanne Simpson, a Nishaabeg scholar and activist from Turtle Island[7] speaks to this need:

---

[6] There are many sources that address these ideas. One particularly interesting and provocative reference is Leanne Betasamosake Simpson's, *As We Have Always Done: Indigenous Freedom Through Radical Resistance.*

[7] Turtle Island is the name that many Indigenous nations use for the continent that is colonised as North America.

It is most critical for Indigenous Peoples and our allies to discuss good relationships in terms of alliances and solidarity in times of relative peace, when we all have time to retreat, re-evaluate, challenge, reflect, and envision. When we have the space to consider how to interact with each other in a respectful, responsible way – in a way that promotes the kind of justice that we are seeking on a grander scale, one that honours the very best of our traditions."[8]

Yet, many Settlers lack the knowledge and experiences that are foundational to the kinds of cross-cultural relationship-building that Simpson calls for. In our experiences of opening conversations with post-secondary students about colonialism and the historical and ongoing oppression of Indigenous peoples – in particular, the history of Indian Residential Schools in Canada – we often hear reactions of "why didn't we learn about this before now?" Indeed, for those who live outside of the experience of Indigenous peoples, and even knowledge of their ongoing existence, it can be a surprise to learn both that Indigenous cultures continue to flourish around the world, and that they often do so in the face of a long and continuing history of oppression by colonial cultures. These students' shock comes as an otherwise invisible veil is pulled away from the dominant reality in which they comfortably live. Miq'maq theorist Marie Battiste describes this Eurocentric state of being as an "enforced cultural imperialism on Aboriginal knowledge and peoples".[9] For her, it is perhaps unsurprising that Indigenous existence is largely or entirely erased in the colonised cultural mindset because "Eurocentrism is the dominant consciousness and order of contemporary life. It is a consciousness in which all of us have been marinated."[10] While Battiste's conception of Eurocentrism is a reality experienced people around the world, for both Indigenous and non-Indigenous peoples, it is not a fixed or unavoidable state. Rather it is a set of socially engineered relations that can be unmade if the moral, social, and political will can be summoned. Indeed, environmental educators can be a catalyst in this summoning.

Metis scholar Gregory Lowan-Trudeau notes that there are "an increasing number of scholars and educators who advocate for the integration of Indigenous, Western, and other knowledges in our collective attempts to address the world's current ecological crises."[11] Such integrations, however, require educators who have a basic consciousness of colonial legacies in their regional, national, and global contexts. Realising Battiste's colonial marinade metaphor, how can educators and learners spark an awakening that allow for shaking off our shared colonial immersion?

---

[8] Leanne Simpson, "First words," xiv.
[9] Marie Battiste, "You can't be the global doctor if you're the colonial disease," 126.
[10] Ibid., 124.
[11] Gregory Lowan-Trudeau, *From bricolage to metissage*, 5.

The following activity is designed with such an awakening in mind; or, at least a gesture toward the potential for helping educators, but especially non-Indigenous educators, to develop the range of consciousness, sensitivities, and dispositions to help others to understand the importance of working collaboratively for right relations that can counteract colonial histories.

## LEARNING ACTIVITY 6.4: WORLD MAP

1. Find an open space outdoors or clear away the furniture from an indoor learning space to create an open area in which participants can move around without constriction.

2. Place any kind of small visual marker on the ground in the centre of your space (a water bottle, or a ball or bean bag work well). Explain to your group that together you are going to imagine a world map unfolding underfoot. The marker in the centre of the space is representative of the place on the earth that you all currently occupy – your learning space, in your community, in your town/city, in your region, in your nation, in the world. Moving out from the space marked on the imaginary map is the rest of the world expanding in all four cardinal directions (north, south, east, west). It is helpful to identify which direction is north, south, east, and west to your location.

3. Let students know that you are going to be discussing movements of humans around the globe over centuries, and the colonial relations that have emerged as a result of those migrations. Talking about colonialism can be emotional for some people, especially for Indigenous people whose land has been colonised, and others who have close ties or proximity to colonial conflict. Attempt to create invitations that honour these emotions and enable respectful dialogue. During the activity, watch participants carefully to gauge reactions. After the activity, take time to check-in with any participants who may seem disturbed or uncomfortable.

4. Explain that the activity will proceed by the leader calling out a prompt that will help participants to move to a location on the imaginary map. Also explain that because the group is inventing the map based only on your current location, and the four cardinal directions, that the scale of the world map may be unusual, and may shift and change over the course of the activity. Explain that between the various prompts there may be opportunities to stop and discuss peoples' choices of where to locate themselves on the map. It may be helpful to provide an example so that participants understand what is being asked of them, for instance the instructor may say, "If our marker represents where we all stand today, and I want to move myself on the map to India, then

I would place myself in *here* on the map because India is south-east of our current location."

5. Provide some or all the following prompts and allow participants to place themselves on the map for each one. Note that the prompts are written in order from what is expected to be most innocuous to the more contentious or thought provoking in relation to colonialism and environmental ethics. However, every group is different and the leader will need to judge which prompts to give in which order based on their groups' needs. Leaders are also encouraged to develop their own prompts tailored to the context in which they are leading.

- Move to a place on the map where your favourite international cuisine comes from.
- Move to a place on the map where you might like to visit or go on vacation.
- Move to the place on the map where your favourite film, novel or story takes place.
- Move to the place on the map where your ancestors were born.
- Move to the place on the map where you were born.
- Move to the place on the map where you live now.
- Move to a place on the map where you know about a land rights conflict between Indigenous peoples and Settlers.
- Move to a place on the map where you know of a grievous social injustice perpetrated by a government on Indigenous, or other groups of, people.
- Move to a place on the map where you know there has been a historic or contemporary environmental disaster that impacts Indigenous peoples, Settlers, or both?
- Move to a place on the map where you know of a success story in terms of relations amongst Indigenous peoples, Settlers, and environmental ethics?

6. Debrief: Ask participants "what did you notice about our individual and collective movements around the imaginary map as the activity progressed? What do these movements mean for the ethical relations that exist between Indigenous peoples in a given area and Settlers who migrate?"

7. Leadership Notes: Depending on how much time is available, stop between some prompts to discuss where participants have placed themselves and why they selected those locations. This may take the form of small group or partner talk amongst participants standing nearest each other on the "map", or a whole group survey conducted by the leader. For some prompts, it may be appropriate to ask participants to discuss the ethical significance of their selection of position, or movement on the map, e.g.: Did you notice a mass movement in that round? What might that mean for Indigenous peoples, Settlers, and

the land they occupy? Note that the earlier prompts are intended as "warm-ups" that can set a tone for sharing more contentious or personal ideas and values in the later prompts, but these "warm-ups" are not necessarily disconnected from issues of environmental ethics or Settler-colonialism, for instance international cuisine and vacation travel are often associated with financial privilege that may be more associated with groups of people who benefit from colonialism in ways that colonised peoples do not.

All these issues are ripe for discussion. In groups where some participants identify Indigenous ancestry in the area of the world around your "here" marker, those people may be willing to offer some perspective from their context – though, be cautious about tokenising or asking individuals to speak for a whole group. In situations where no Indigenous representation exists, the leader may choose to find strategies to bring the voice of Indigenous peoples into the exercise through Indigenous writing, poems, song, artwork, etc. Finally, this activity has been conceptualised and workshopped in a North American context where the primary colonial history is of migration from Western Europe. In workshops we have found this activity to provide a poignant visualization of that migration. Leading the activity with participants whose experience of Settler-colonialism is different from the North American context will likely result in different outcomes. We would be pleased to hear about how the activity illustrates colonial patterns in a broader global context.

## EXTENSION ACTIVITY

Working in small groups, use a paper-based or digital world map to plot some of the movements that you noticed during the virtual world map exercise. What is interesting or significant about these patterns of movement? How do maps help us tell stories that explore ethical relationships among Indigenous peoples, Settlers, and land?

## SOME THOUGHTS ABOUT THEORY

This chapter's heading poses the question: "Is there a place for my history, my culture, my feelings?" The activities and orientating ideas shared so far in the chapter indicate that it is, in fact, not possible to contemplate ethical action-taking in a vacuum as if historical social and cultural practices do not influence what we care about and how we respond! This closing section of the chapter shares some useful ideas about:

- relational ethics, and
- the power of cultural artefacts (metaphors, images, discourses).

These may be useful in developing your own understandings and responses to ethical deliberation and action.

## Relational ethics

When we recognise the place of history, culture, materiality, social circumstances, and feelings in environmental matters, we inevitably find ourselves working within a relational understanding of the world. Relationality recognises that human agency, learning, and social change develop through myriad interactions of mind-body, past-present-future, individual-collective, social-ecological, powerful-less powerful, and so on.

This chapter has shared several examples of the relationality of people's ethical deliberations and actions. Recall how Martine's decision not to report illegal fishing was made *in relation to* her compassion for the financially needy fishing community, *in relation to* her love and respect for the integrity of marine and coastal ecosystems and the area she calls home, *in relation to* her sense of moral obligation to future generations, *in relation to* her professional obligations in upholding laws of the country, and so on. Similarly, Nkanyiso's moral struggle with vegetarianism unfolded *in relation to* his family bonds, his Zulu heritage, his developing identity as an environmentalist, his dietary preferences, peer pressure, and a developing moral commitment to a sustainable future to people and planet. More than just adhering to a set of values when deciding whether to stop eating meat, Nkanyiso is guided by his relationships to people, places, and things past, present and future.

The unpredictable interaction of these and other relational dynamics, each in their unique combination of circumstances, are what make Martine's and Nkanyiso's, as well as all our ethical deliberations so contextually rich and open-ended. So, when we speak of the "embodied" nature of ethics,[12] we acknowledge that our decisions to act in certain ways because of the things we care about are grounded in the landscapes we inhabit(ed), the spaces we pass(ed) through, the meals we ingest(ed), the comforts and discomforts we experience(d). We are also called upon to consider what it means to act well *in relation to* the networks of friends, family, ancestors, neighbourhoods, colleagues, and the plants and animals that we nurture, kill, attract or repel in order to be fed, safe, entertained or comforted.

When we speak of ethics as being "contextual and contingent",[13] we acknowledge that our ethical choices and actions are situated within these dynamic relationships and stories that are not time-bound or space-bound. It is for these reasons that we can describe our ethical engagement

---

[12] See, for example, Sauvé, "Being Here Together," 2009.
[13] See, for example, Lausanne Olvitt, "Doing What's Right for People and Planet"; Lausanne Olvitt, "Working with environmental ethics and adult education."

in the world as being historically and socio-culturally emergent, that is, our ethical actions *emerge* out of the unpredictable and unique relationships our life's circumstances. A relational orientation to environmental ethics is well-captured by Catherine Roach who states: "To be human is to be in relationship. We are constituted and shaped by our relations with others that occur both in external reality and inner psychic life."[14]

Numerous traditions and positions within environmental philosophy are based on these kinds of relational understandings of ourselves in the world such as deep ecology, ecofeminism, social ecology, and bioregionalism. The relationality inherent in deep ecology processes and Arne Næss' personal philosophy of Ecosophy-T have been discussed in the previous Chapter 5 and are again in Chapter 8.

In this chapter, we introduce a different relational philosophy from southern Africa known as *Ubuntu* that, although not conventionally recognised as an eco-philosophy, still offers some important considerations about how we might live in the world – and even the cosmos.

*Ubuntu* is about "humanness" and is often associated with the proverb *Umuntu ngumuntu ngabantu* – a person is a person through other people. Conventionally, this has reinforced the view that *Ubuntu* is a people-centred, anthropocentric philosophy with limited relevance to environmental values and ethics. Recent scholars, however, point out that *Ubuntu* is a philosophical thread of African epistemology that is relational and recognises that human life and well-being is embedded in the well-being of the rest of the natural world. People and their actions are thus related to natural entities, and to past and future generations. Recent African environmental philosophers argue, therefore, for a fuller understanding of *Ubuntu* that recognises human connectedness as a reflection of relatedness to the entire biophysical world and cosmos.[15] Le Grange[16] explains that "an anthropocentric reading of *ubuntu* is flawed and that by definition it means relatedness to (or embeddedness in) the web of life."

A relational view of the world, whether through *Ubuntu*, deep ecologies, eco-feminism, bioregionalism and so on, keeps us gently attuned to the awareness that our lives – even to the smallest detail, are the steadily growing manifestations of relational webs that cross time, space, culture, and emotion. We can contemplate, for example, whether the meal we are about to eat has been produced within local nutrient cycles and seasonal rhythms, or produced through pesticide and inorganic fertiliser-induced economic cycles and the delivery schedules of long-distance refrigeration

---

[14] Catherine Roach, *Mother/Nature: Popular culture and environmental ethics*.
[15] See, for example: Munyaradzi Murove, "An African Commitment to Ecological Conservation," 195-215; Lesley Le Grange, "*Ubuntu, Ukama* and the Healing of Nature, Self and Society," 56 -67; Lesley Le Grange, "*Ubuntu/Botho* as Ecophilosophy and Ecosophy," 301-308; Mogobe Ramose, "Ecology through *Ubuntu*," 69-76.
[16] Lesley Le Grange, "*Ubuntu, Ukama* and the Healing of Nature, Self and Society," 307.

trucks. Awareness of these things, and the decisions we make in relation to them, are shaped by (amongst others):

- **Our personal socio-economic circumstances** – Can we afford organically grown food, or have the time and physical means to grow our own?
- **Historic-geographical factors** – Do we live in a region where food can be cultivated easily, and what kinds of food production systems and networks have been established in this place over the years?
- **Cultural influences** – Why do we eat the foods that we do, in the forms and quantities that we do?

## The power of cultural artefacts

Val Plumwood[17] places a strong emphasis on the cultural influences of ethical living. She goes so far as to state that the root cause of the global ecological crisis is cultural. Here, she is not referring to specific cultural traditions or regional groups of people with distinctive styles of music, cooking or clothing; she is referring to culture in the broader sense of how societies understand and conduct themselves in relation to Nature and the rest of the world. As such, the ethical attentiveness that Plumwood calls for is a relational ethics that draws attention to our (often dysfunctional) relationships with the natural world. Plumwood describes the world's dominant culture as having elevated human rational thought above emotion and embodiment – a dominant culture that separates the world dualistically into reason versus nature, mind versus body, progress versus primitivism and so on. The result is a world in which non-human nature is seen as a commodity to be dominated and exploited.

Plumwood suggests that the only way to transform this dominant global culture of mastery and exploitation is to disrupt the cultural view that the natural world is external and passive, and to recognise and reconnect with our material and ecological support base with sensitivity, humility, and generosity. However, this kind of profound cultural change is extremely hard to achieve because our economic, political, and educational systems are interwoven so tightly with one another and with the dominant rationalist culture of domination that even our smallest daily routines are reflections of the dominant narrative. In earlier work, Plumwood describes it thus: "The strands interwoven by this master story of colonisation form a mesh so strong, so finely knit and familiar it could almost pass for our own bodies, but it is an imprisoning web which encloses us."[18]

The following sections on:

- cultural images and metaphors, and

---

[17] Val Plumwood, *Environmental culture: The ecological crisis of reason.*
[18] Val Plumwood, *Feminism and the mastery of nature*, 195.

- discourses foreground just a few of the many ways that our backgrounds, cultures, and social circumstances shape our ethical responses to a planet in crisis. These short introductions are not comprehensive but aim to springboard your further reflections and transformative actions.

## Cultural images and metaphors

It is widely recognised that the images and role models circulating in society massively influence our values, priorities, aspirations, and action-choices. Advertisers and other propagandists work to exploit this human feature as much as possible. Let us optimistically re-imagine what kinds of cultural artefacts – images, metaphors, products and vocabulary – might reflect a powerful cultural shift away from the narratives of mastery and exploitation that Plumwood critiques. What if the instruction manual accompanying a new electronic device were to give a transparent account of the social and environmental impact of its production process and give owners guidelines for keeping the device for as long as possible, as well as contact details of depots for responsible recycling of the device after its useful life? This would shift the dominant cultural image of electronic devices from being consumer and status-driven commodities, to being important but environmentally costly assets whose durability and long service is a marketable feature. For a more in-depth exploration of these ideas, you could re-read the section "Being Critical" in Chapter 3.

## Discourses

Discourses are the shared ways that groups of people think, talk, and write about certain topics. Quite a lot of social power and influence results from discourses reflecting shared ideas. When they circulate in society, it can be difficult not to be inspired, intimidated, angered, or validated by them. For example, we can identify a capitalist discourse, a religious extremist discourse, a Marxist discourse, a racist discourse, a feminist discourse and so on by reading or listening carefully to the types of words and metaphors used, by noticing the grammatical structures, and the attitudes or perspectives they convey. Discourses hold power in society because, by the time they are recognised as discourses, they are already endorsed by large numbers of people and have a social momentum that sustains them and the values they reflect.

Due to the diversity of environmental values and the contestation surrounding almost all environmental issues, there is no singular environmental discourse, only multiple environmental discourses. Environmental discourses can be subtle and pervasive: look out for them in newspapers, on websites, in school textbooks, in advertisements, in policy documents, institutional vision and mission statements – and in this book you are reading now!

As educators, it is important for us to be alert to discourses and the way they can influence our understanding of and responses to people-nature relations. As reflexive citizens, we need skills to recognise the range of environmental – and other – discourses circulating in society so that we can ask important questions such as: What values are evident in this discourse? Where and, importantly, why did this discourse develop? What kinds of ethical responses to social-ecological concerns are encouraged through this discourse? In whose interests is it to sustain this discourse? What is my personal stance in relation to this discourse – and why? Does this discourse play out in my educational work – and to what effect?

This chapter has foregrounded that we are more than just "products" of the social world we inhabit, as radical constructivists would have us believe! The very fact that we experience feelings and can be reflexive and future-oriented as we deliberate our options establishes us as moral agents. To this end, Chapter 8 stimulates further thinking about the ways that ethics are caught up in our everyday practices, and Chapter 9 offers some starting points for re-imagining the present and trying out alternatives that reflect a different or expanded ethics.

## References

Bauman, Zygmunt. *Postmodern ethics*. Oxford: Blackwell Publishers, 1993.

Battiste, Marie. "You can't be the global doctor if you're the colonial disease." in *Teaching as activism: Equity meets environmentalism*, eds. Peggy Tripp and Linda Muzzin. Montreal, PQ: McGill-Queens University Press, 2005, 121-133.

Le Grange, Lesley. "*Ubuntu, Ukama* and the Healing of Nature, Self and Society." *Educational Philosophy and Theory* 44, no. S2 (2012): 56-67. doi: 10.1111/j.1469-5812.2011.00795.x

Le Grange, Lesley. "*Ubuntu/Botho* as Ecophilosophy and Ecosophy." *Journal of Human Ecology* 49, no. 3 (2015): 301-308. https://doi.org/10.1080/09709274.2015.11906849

Lowan-Trudeau, Gregory. *From bricolage to metissage: (Re)thinking intercultural approaches to Indigenous environmental education and research*. New York: Peter Lang, 2015. https://doi.org/10.3726/978-1-4539-1527-1

Murove, Munyaradzi. "An African Commitment to Ecological Conservation: The Shona concepts of *Ukama* and *Ubuntu*." *Mankind Quarterly* 45, no. 2 (2004): 195-215. https://doi.org/10.46469/mq.2004.45.2.3

Olvitt, Lausanne. "Working with environmental ethics and adult education: Some experiments and reflections from southern Africa," in *Learning in a changing world: Selected papers from The 4th World Environmental*

*Education Congress*, eds. Lausanne Olvitt, Linda Downsborough, and Heila Sisitka. Howick, South Africa: EEASA, 2009, 43-45.

Olvitt, Lausanne. "Doing What's Right for People and Planet: An investigation of the ethics-oriented learning of novice environmental educators." PhD dissertation, Rhodes University, Grahamstown, South Africa, 2012.

Plumwood, Val. *Feminism and the mastery of nature.* New York: Routledge, 1993.

Plumwood, Val. *Environmental culture: The ecological crisis of reason.* London, UK: Routledge, 2002.

Ramose, Mogobe. "Ecology through *Ubuntu*," in *Environmental Values Emerging from Cultures and Religions of the ASEAN Region*, ed. Roman Meinhold. Bangkok: Konrad-Adenauer-Stiftung & Guna Chakra Research Center, 2009, 69-76.

Roach, Catherine. *Mother/ Nature: Popular culture and environmental ethics.* Bloomington: Indiana University Press, `2003.

Sauvé, Lucie. "Being Here Together," in *Fields of Green: Restorying culture, environment, and education*, eds. Marcia McKenzie, Paul Hart, Heesoon Bai and Bob Jickling. Cresskill: Hampton Press, 2010, 325-335.

Simpson, Leanne. "First words," in *Alliances: Re/Envisioning Indigenous-non-Indigenous relationships*, ed. Lynne Davis. Toronto: University of Toronto Press, 2010, xiii-xiv.

Simpson, Leanne B. *As We Have Always Done: Indigenous Freedom Through Radical Resistance.* Minneapolis, MN: University of Minnesota Press, 2017. https://doi.org/10.5749/j.ctt1pwt77c

Wals, Arjen. "Learning in a changing world and changing in a learning world: Social learning towards sustainability," in *Learning in a changing world: Selected papers from the 4th world environmental education congress*, eds. Lausanne Olvitt, Linda Downsborough and Heila Sisitka. Howick, South Africa: EEASA, 2009, 43-45.

Weston, Anthony. "Before environmental ethics," in *The Incompleat Eco-Philosopher*. Albany, NY: State University of New York. 2009, 23-43.

# CHAPTER 7

## ISN'T ALL KNOWLEDGE ETHICS-BASED?

**Pedagogical intent:** This chapter challenges the notion that knowledge (that animals feel pain, for example) leads to the formulation of ethics theories and resulting prescriptions for action. On the contrary, Jim Cheney and Anthony Weston suggest that ethics are what lead us. How we carry ourselves will determine how we see and frame the world. For these authors, environmental etiquette becomes an *a priori* concern. Our actions thus shape the way we see the world and determine what we learn about the world.

**Main activity:**          7.1: Case studies of ethics and etiquette

**Main theoretical resources:** Jim Cheney and Anthony Weston's theorising about ethics-based epistemology is the primary resource. Aldo Leopold supplies a powerful supporting example.

## Ethics-based epistemology

There are many ways to view the world. How we see it depends on how we move through our cultural and natural landscapes, and what we find ourselves seeing. Jim Cheney and Anthony Weston talk about this approach to the world as etiquette.[1] Etiquette is generally thought of as an ethics of practice, but these two philosophers take the term a little further. For them, the way we carry ourselves in the world will affect what we will learn. They suggest that if the conventional etiquette for learning is shifted, such that we see the world in a different way, we will learn new things and see things that have been overlooked or under-valued. This study of the nature of learning and knowledge is called epistemology. This work is epistemological in that they are examining relationships between values – or ethics – and knowledge creation.

Our daily ethics – that is our etiquette towards the world – will shape what we learn; it will influence our ability to accept or reject certain knowledge claims. That is our epistemology. Simply, ethics shape our epistemology. This, of course, occurs in our daily interactions with the world – how we interpret the news, respond to advertisements, and choose recreational activities. However, this etiquette also has profound effects on how research is conducted. Consider the following example.

Quetico Park is in Northwestern Ontario, Canada. It is a "wilderness area" and home to the Lac La Croix First Nation. The evolution of the park, and its shared management with the Aboriginal inhabitants, has

---

[1]    Jim Cheney and Anthony Weston, "Environmental ethics as environmental etiquette," 115-134.

been slow and is ongoing. However, in recent years, much interesting progress is being made. One group on the forefront of this progress are the archaeologists working in the park. They recognise that practices in their field, or the etiquette of traditional archaeology, have typically ignored the viewpoints of Aboriginal peoples. The result is that knowledge of the past has been constructed through a European filter.

In beginning to work collaboratively with the Lac La Croix First Nation, the researchers learned that the Elders did not make a distinction between past and present. For them, all the proposed research sites were contemporary sites. That is, they are still very much in use through stories, songs, memories, and religious practice. The spirits that inhabit this landscape make it alive in a different sense than seen by most Euro-Canadians. This aliveness of the landscape means that things – what archaeologists call artefacts – must be left in place for their continued use. It also means that they must be shown respect. Here, too, respect looks a bit different than the way that it is often conceived. Elders requested that offerings of tobacco be placed on the sites where artefacts are found. Here, respect is not conceptual – it is not found in words – it is an act. These acts of gift-giving are integral parts of an etiquette that has grown out of a particular landscape.[2]

It is interesting to see how a new research etiquette can open new windows to the landscape. The working out of new research protocols are described in the following quotation:

> The first priority of the Elders was to protect the integrity and spiritual nature of the sites and the artefacts they contain. They did not exclude research as long as the guiding principle of respect was upheld. As a sign of respect, they requested that tobacco be placed on sites where artefacts are found. The belief in "a spiritual landscape within a physical landscape" led to two fundamental changes in the way research in Quetico Park was conducted: 1) Surveys were conducted so that no disturbances occur to the site. Consequently, no shovel tests, test pits or excavations were allowed. 2) All artefacts were left on the site where they were found. The objective of the archaeological survey shifted from "collecting" to "recording."
>
> Initially, this "no collecting" policy was seen as impeding archaeological research in the park, but we found that this is clearly not the case. The standard archaeological procedures of drawing site maps, and describing and photographing artefacts were still followed, The Elders' restrictions on not digging or otherwise disturbing the site were followed, and information about artefacts was collected in the field, but the artefacts themselves were left at the site. This system of finding artefacts and leaving them on the site became affectionately known as "catch and release archaeology."

---

2    Jon Nelson, Quetico: Near to Nature's Heart.

1995 was Quetico's year of fire; more acres burned that summer than in the previous sixty years combined. These large, intense forest fires greatly increased ground visibility and provided a superb opportunity to complete an archaeological survey while working within the guidelines developed with the Lac La Croix Elders. Andrew Hinshelwood's 1996 post-fire survey, centred on Kawnipi Lake, found that ceremonial activities were evident in the pattern and nature of artefacts present at some of the sites, lending considerable objective support to the Elders' statements about the spiritual nature of the sites.

In 1997, Frank Jordan (from the Lac La Croix First Nation) and I conducted archaeological surveys in other areas of the park that had burned in 1995. Although it was the second summer after the fire, we were still able to take advantage of the resulting increased ground visibility. The following two summers, we, along with students from Lac La Croix and Atikokan, searched the shorelines of French Lake, Pickerel Lake, Rawn Lake, and Batchewung Lake. These lakes – we refer to them as the Pickerel Lake complex – can be considered as one water body since no portages, even in low water conditions, are required to go from one lake to another. Although the advantages created by the 1995 fire were minimal after three years of regrowth, the very low water levels in 1998 made it much easier to find artefacts along the shoreline. All of these assessments found evidence of an extensive occupation from the immediate-post glacial times through to the fur-trade era.

In 1988, prior to the agreement with the Elders, I led a team that searched the Canadian side of the Knife Lake to determine the location of quarries used to obtain Knife Lake siltstone, a lithic material used in the making of stone tools. Two subsequent events, an extensive blowdown that occurred during the storm of July 4, 1999, and the subsequent prescribed burn on the Canadian side of the Lake in October 2000, created ideal conditions for conducting a survey in the summer of 2001 using the Elders' guidelines. The diminished ground vegetation led to the discovery of new quarry sites and four workshops (sites where stone tools were made). During both surveys, I worked with crews from Atikokan and the Lac La Croix First Nation.

Although the guidelines were worked out with the Elders did not allow diggings in the surveys conducted from 1996 to 2001, we found that much more ground could be covered since time-consuming activities such as shovel-testing were not allowed. We were able to investigate places such as shorelines and islands without any campsites that are not usually examined in archaeological surveys. This resulted in the discovery of many unexpected, interesting sites that we otherwise would not have found.[3]

---

[3]  Ibid., 84-85.

# HANDOUT: WOLF STORIES: REFLECTIONS ON SCIENCE, ETHICS, AND EPISTEMOLOGY

Killing wolves to benefit humans is part of contemporary Canadian history. For example, the first biologist hired for Canada's Yukon Territory, in the 1950s, was assigned the task of overseeing a widespread wolf-poisoning program. "Wolf management" remains controversial in Canada.

In 1997, the Yukon Government completed a five-year control program that entailed the killing of 80% of the wolves in a 20,000 square kilometer region of southwest Yukon. The killing was accomplished by shooting wolves from helicopters or, in later years, strangling them in snares. The justification for culling was concern for declining numbers of caribou in particularly vulnerable herds. In this case, the objective was to prevent the Aishihik Caribou Herd from falling below a level where a natural recovery would not occur quickly enough. Because caribou are important to First Nations for subsistence hunting, and because other hunters in the area wanted their so-called "fair share," the Government of the day judged that a long-term natural recovery, estimated to be between twenty and thirty years, would not be acceptable to local people.

Not everyone supports wolf kills and many spoke out in opposition. Some were concerned that too little was known about why this herd was declining. Others claimed the wolf was being made the scapegoat for past excesses and that over-hunting was responsible for the problem. For many, the critical question was whether the wolf kill was a reaction to a biological problem or a treatment for the symptoms of a much deeper human problem.

In the later stages of the wolf-control program, biologists began experimenting with wolf sterilization techniques as an alternative to shooting. The method involves: (a) tranquilizing what are thought to be alpha wolves from a helicopter, (b) moving the wolves to a nearby community and placing them in a holding crate, (c) on arrival of a veterinarian, anesthetising the wolves then sterilizing them with either a tubal ligation or a vasectomy, (d) placing wolves in release boxes for the night, and, (e) then returning them to the capture site the next day. In local newspapers, biologists promoted this as being less intrusive, and that "sterilization will reduce the frequency in which aerial hunting must be carried out. That will soften public resistance and be less of a drain on financial resources. This takes us deep into the ethics of science.

Science only provides snapshots of reality; much remains unknown and uncertain. Often our ethical choices are framed and driven by what we know. In the Yukon example, biologists and managers are choosing to frame the discussion in terms of what they know – wolf control practices that are either "more" or "less" intrusive.

It is important to understand how, intentionally or implicitly, science can limit available ethical choices, and how science rests on the working values and assumptions of practitioners.

## HANDOUT: CONTINUED

For example, in planning for the Yukon wolf kill, one biologist said: "There are two options: intensively manage or let the caribou population follow a natural decline. Both are defendable." That is, scientists can intervene in the ecosystem by removing wolves and monitor the effects, or allow the system to decline and/or recover without human intervention and again monitor the effects. The same biologist went on to say that the "Public in Yukon prefers intensive management on this herd." A colleague added, "We need to address public policy . . . and design something scientific, with a set of hypotheses and alternate hypotheses." At this point, science and ethics are inseparable, and we begin to be defined by what we are prepared to find out, and hence by what we know. Ultimately, research questions are rooted in our values – and our ethics – making all knowledge value-loaded and ethics-based. And, when we operate from different value systems, we learn different things and we tell different stories.

We find it difficult to imagine how a helicopter-assisted killing can be considered respectful, or how the sterilization of dominant pairs in a wolf pack is considerate. People, particularly scientists, are not eager to talk about this. However, Tlingit Elder Harry Morris, commenting at a meeting before the wolf kill, provides those willing to listen with much to think about. He reminds us that his people have, at times, killed wolves, that it was part of their culture, "but the wolf must not be made a fool of." For Mr. Morris, right relationships and right conduct matter.

What scientists do matters; it counts ethically. A good starting point for scientists would be to preface all reports with a discussion about the ethics of the project. What considerations, for example, guided the choice of research questions? How were research practices mindful of entities in the natural world? Let us explore, each time out, our deeper positions – our foundations. This is not a fanciful dream; it is being done now. Revealing considerations make them part of the on-going conservation conversation.

This reminds us that research assumptions must be considered, supported, and occasionally revised, but never taken for granted. Such discussions also help to prepare advocates to participate in the processes that occur whenever conservation proposals begin to live in a context. We challenge scientists and managers to place everything they do into the context of ethics and then spend 25% of their time thinking and writing about the ethical dimensions of their work – on every project.

In the end, if we are content to train a society that is expert in its ability to dig up, chop down, and shoot dead (or sterilize) the world around it, then that is how public discussion will be framed. That is what our professionals will do, and that will define how our society will be realized. Surely it is time to enact a different story.[4]

---

[4]   This text consists of edited selections from the paper: Bob Jickling and Paul Paquet, "Wolf Stories," 115-134.

## LEARNING ACTIVITY 7.1: CASE STUDIES OF ETHICS AND ETIQUETTE

1.  Carefully read the block quotation about the Lac La Croix First Nation as told by Jon Nelson in *Quetico: Near to Nature's Heart*. In what ways were the researchers required to demonstrate an etiquette different to that normally practiced in archaeology? How did it enable them to see the world differently? Imagine other kinds of research that could yield interesting results if they were consciously ethics-based, and thus grounded in a different etiquette.

2.  Read the handout, "Wolf stories." What assumptions do these scientists hold about the work they do? How do they seem to see their relationships with wolves? Based on their actions in this example, describe the scientist/nature relationship. What kinds of things might scientists learn if they changed their etiquette? Imagine a few possibilities. Would you like to see scientists discussing the ethics of their research at the beginning of each report? Why?

## EXTENSION ACTIVITY

3.  Find examples of research reporting in the popular press. What etiquette scaffolds the research reported? How does it shape the results? What questions would you like to see posed? What would you like to ask the researchers?

4.  Design your own nature-based research project. What kind of etiquette would you like to pursue? Think carefully about your assumptions. What ethical, sociological, and research assumptions would you propose? What would be the advantages of your approach? The disadvantages?

## SOME THOUGHTS ABOUT THEORY

*Environmental Ethics as Environmental Etiquette: Towards an Ethics-Based Epistemology*

Jim Cheney and Anthony Weston bring yet another radical turn to the unfolding of critical perspectives and possibilities in environmental ethics in their paper called "Environmental Ethics as Environmental Etiquette: Toward an Ethics-Based Epistemology".[5] Like Arne Næss (see Chapter 5), they take a critical look at substantive and methodological assumptions about the nature of ethics itself. While Næss was sceptical about the

---

[5]  Jim Cheney and Anthony Weston, "Environmental ethics as environmental etiquette," 115-134.

usefulness of deontological ethics, or ethical reliance on duties to guide behaviour, Cheney and Weston are concerned about the relationship between knowledge and ethics, hence epistemology. They begin their argument by observing that ethics typically arises out of knowledge of the world – the facts. They then pose the question: What if it is the other way around? Suppose that knowledge actually arises out of our ethical practice – our comportment, our etiquette, or the way we inhabit the world?

To make their point, Cheney and Weston compare traditional perceptions of relationships between ethics and epistemology within the field of environmental ethics with alternative possibilities. For example, in what they call epistemology-based ethics – or the traditional view within the field – ethical action is a response to knowledge of the world. That is, responses are based on the knowledge that animals – mammals at least – are conscious of their own welfare and are goal directed. Based on this knowledge, it is inhumane to treat them as a means to a human end. Or, because animals feel pain, it would be wrong to cause them suffering. If we hold, then, that it is our duty to minimise pain and suffering, or to treat animals as having rights to realise their own ends, then we are pursuing deontological ethics. This approach places epistemological considerations first.

In countering an epistemological based ethics, Cheney and Weston begin by revealing assumptions that underlie this approach. If epistemology comes first then it is assumed that the world is, indeed, knowable. From this follows a second assumption that is, as more is known, ethics can be extended accordingly, and incrementally. Perhaps this is what can be said of Taylor's theory of respect for nature, discussed in Chapter 4. Working from a more solid core of understanding about sentient beings and subjects-of-a-life, he has proposed expanding ethics more broadly to teleological centres of life, such as plants. From here it follows, as a third assumption, that the task of ethics is to make sense of the world – to figure out what matters, ethically. What warrants moral considerability? How far can ethics be extended?

It becomes an interesting proposition to begin reversing these assumptions. How would environmental ethics be challenged if, for example, we assumed that what can be known is value-driven? Cheney and Weston ask us to consider how knowledge of the world is determined by how we approach it – by the values that are reflected in our etiquette. Then they ask, how the world might look differently if we approached it differently – with different ethics, by carrying ourselves into the world differently. If we were to take this idea seriously, then an *a priori* concern for all researchers would be to consider how they ought to approach the world in their pursuit of knowledge. What, for example, would be an ethical approach to anthropology? To natural and environmental science? This is a radical shift for research, and for ethics, too. Instead of looking for knowledge claims to frame ethical discourse, this means that ethics are primary; they open the way to knowledge.

Using examples given in this chapter's activities, we can explore these ideas a little further. For example, when archaeologists begin their work with a different kind of respect for artefacts – or a different respect for the whole landscape, including the artefacts present – they are freed from pre-existing assumptions. When they no longer collect and remove artefacts, and they no longer dig up the land, they can see this landscape anew. They see and find things that they have not noticed before. Similarly, the results of the wolf experiments described in the "wolf stories" can reveal some knowledge about wolves when subjected to harsh treatment conditions during population manipulations – that is vexed wolf populations. However, these stories do not tell us about a wolf population that is unfettered (or much less fettered) by human manipulations; in other words, how wolves respond to their environments when they are not made a "fool of." What, for example, do they need to thrive? How do they move through their territories? Do their populations eventually self-regulate within their ranges?

In posing concern for ethics, in general we are trying to discover what things in the world demand practical respect. Cheney and Weston challenge us to think about understanding how the world would be different if ethics came first – if ethics guided or framed our inquiries. Looking more systematically at the assumptions that underlie ethics-based epistemology, they begin by suggesting that ethical action is about opening possibilities to enrich the world. When considering animals, for example, we need not follow the usual pattern of finding out what they are capable of before deciding how to consider them ethically. Rather they state, "we will have no idea of what other animals are capable of – we will not really understand them – until we have already approached them ethically."[6]

Second, instead of assuming that the world is ultimately knowable, Cheney and Weston suggest that knowledge of the world has barely begun to unfold for us – "hidden possibilities surround us at all times".[7] From here they assume that knowing the world is not systematically incremental, and hence ethics should not be thought of as extensionist. Rather, ethics can be discontinuous, but also dissonant and pluralist. Ethical discoveries are always possible, and sometimes disruptive. Finally, when approaching the world from an ethics-based epistemology perspective, the task of ethics is to enrich the world and our knowledge of it. Rather than sorting the word into the morally considerable haves, and the have nots, the task is more inclusive and open ended.

---

[6]    Ibid., 118.
[7]    Ibid., 118.

Building on the work of Tom Birch,[8] Cheney and Weston suggest that ethics-based epistemology begins with consideration of everything, insofar as we can. Birch calls this "universal consideration" whereby "others are now taken as valuable, even though we may not yet know how or why, until they are proved otherwise."[9] Such a view carries obligations. As Cheney and Weston argue, "universal consideration requires us not merely to extend this kind of benefit of the doubt but actively to take up the case, so to speak, for beings so far excluded or devalued."[10] Fundamental to this requirement will be seeking to understand how we ought to approach those things that we respect. Ethics – that is etiquette – comes first.

That we should approach other entities with respect, and that knowledge is ethics-based, is not new. Many traditional ways of knowing have stood in contrast to the epistemological dominance of Western science, including those of Yukon First Nations.[11] Outsiders cannot interpret First Nations cultural traditions; they will have to do that for themselves. Outsiders can only comment on our own experiences while listening to, and working with, Aboriginal colleagues. The legends, stories, and reflections of Yukon First Nations people have placed before us a mirror to reflect upon our own cultural traditions. Of particular interest is the challenge to non-Indigenous cultural frameworks for organizing knowledge – especially the tendency to separate ethical, emotional, and spiritual knowledge from "hard" science.

If all knowledge is ethics-based, and if, in the absence of evidence to the contrary, all entities deserve consideration, then how ought we to approach inquiries about these entities that we respect? First, the idea of universal ethical consideration begins to create new ways of seeing the world. How we create or recreate the world counts. We can no longer be aloof or disinterested observers. Ethics-based epistemologies are concerned with right relationships, and right relationships are grounded in mindfulness. When we are mindful and respectful then we act with courtesy and etiquette, including trans-human etiquette. Louise Profeit-Leblanc, former Native Heritage Advisor, underscores this point when asked to comment on ethics. For her, ethics are "that which we do to ennoble us."[12] Ethics then are more than collections of ideas; they are also performative.

What we do, how we act, and the research procedures we choose all count. Aldo Leopold, often thought of as the father of wildlife management, knew this too. In his famous essay "Thinking Like a Mountain," Leopold

---

8    Tom Birch, "Moral considerability and universal consideration," 313-332.
9    Ibid., 328.
10   Jim Cheney and Anthony Weston, "Environmental ethics as environmental etiquette," 120.
11   See Louise Profeit-Leblanc, "Transferring wisdom through storytelling," 14-19.
12   Profeit-Leblanc, "Transferring wisdom through storytelling," 16.

recounts a life-changing experience that occurred on the day he saw a wolf die:

> We reached the old wolf in time to watch a fierce green fire dying in her eyes. I realized then, and have known ever since, that there was something new to me in those eyes – something known only to her and to the mountain. I was young then, and full of trigger-itch; I thought that because fewer wolves meant more deer, that no wolves would mean hunters' paradise. But after seeing the green fire die, I sensed that neither the wolf nor the mountain agreed with such a view.[13]

For Leopold, shooting that wolf was not ennobling and he felt this deeply – it did not reflect an appropriate etiquette.

## A FEW WORDS ABOUT SCIENCE AND SUBJECTIVITY

In developing this chapter, we are not suggesting that all science is simply subjective and a matter of personal taste and values. As Bruno Latour reminds us, there is a real world out there and science is rather good about helping us to find out about it. Scientific methodologies and practices are also fairly good about providing the needed confidence in their results. When, for example, more than 95% of climate scientists tell us that the earth is heating up and that there will be consequences, we should listen.

What this chapter does introduce, however, is an ethical element into science. Science is not reduced to simply ethics; rather, ethics is a part of science. By being cognisant of the ethical foundations of the questions scientists pursue, and being prepared to broaden those foundations, a richer understanding of the world can become possible. When scientific assumptions are revealed, we can talk about them, and make more informed choices about the type of scientific questions that should be pursued.

Talking about the ethics of science does not make phenomena – like climate change – go away.

### Relativism

Sometimes a tolerance for pluralism is characterised as the beginnings of a slippery slope to relativism. This is not our intention. This chapter is, amongst other things, meant to show how wondrous the world is, and how much more there is to know. That we can shift the nature of what we learn by shifting our etiquette.

By being more conscious of the ethical component in epistemology, we can also be more informed and imaginative about controversial issues.

---

[13]  Aldo Leopold. A *Sand County Almanac*, 138-139.

Rather than slipping into relativistic positions on issues – such as the wolf killing discussed in this chapter – we anticipate the groundwork prepared here can lead to ever more informed and, sometimes intense, discussions about appropriate actions. There will be vigorous, sometimes heated, debate and plenty of disagreement. In the end, value-based positions must be weighed and decisions made.

### Where does this etiquette come from?

When asked, in a private conversation, where this etiquette comes from, Anthony Weston did acknowledge that experience in the world was part of the source. His view, at that time, was that if he had the paper to write again, he might suggest that understanding experiences in the world and emerging etiquette were linked, and that they co-evolve. These important possibilities are discussed in Chapter 5 and again in Chapter 8.

## References

Birch, Tom H. "Moral considerability and universal consideration." *Environmental Ethics* 15, no. 4 (1993): 313-332. https://doi.org/10.5840/enviroethics19931544

Cheney, Jim. & Weston Anthony. "Environmental ethics as environmental etiquette: Toward an ethics-based epistemology." *Environmental Ethics* 21, no. 2 (1999): 115-134. https://doi.org/10.5840/enviroethics199921226

Jickling, Bob & Paquet, Paul. "Wolf Stories: Reflections on Science, Ethics, and Epistemology." *Environmental Ethics* 27, no. 2 (2005): 115-134. https://doi.org/10.5840/enviroethics200527226

Leopold, Aldo. *A Sand County Almanac: With essays on conservation from Round River.* New York: Ballantine, 1966. First published by Oxford University Press in 1949.

Nelson, Jon. *Quetico: Near to Nature's Heart.* Toronto: Dundurn Press, 2009.

Profeit-Leblanc, Louise. "Transferring wisdom through storytelling," in *A colloquium on environment, ethics, and education*, ed. Bob Jickling. Whitehorse: Yukon College, 1996: 14-19.

# CHAPTER 8

## ETHICS IN ACTION, MORAL PROXIMITY, AND CARING RELATIONSHIPS: GOING DEEPER

**Pedagogical intent:** This chapter considers ethics in action as part of everyday activities. Through stories about ethics in action, we explore how ethics arises in and informs everyday practice. These stories will show that transforming modern social practices in ways that are ethically oriented involves a process of struggle against social practices that have come to be normalised. The chapter will provide teachers with guidance on how to analyse ethics as everyday struggle in a complex world.

**Activities:**      8.1: Exploring possibilities for ethics as action

8.2: Chicken and egg

8.3: Zero-waste ethics

8.4: Rhino relations

8.5: Going beyond

**Main theoretical resources:** Zygmunt Bauman's theorising of postmodern ethics, in particular his ethics as proximity thesis.

## Introduction

The importance of moral proximity resonates with other themes in environmental ethics, particularly those introduced in Chapter 5. While they may not arise from the same lineage, they seem to harmonize beautifully. In this chapter we propose to experiment with the overlapping spheres of ethics in action, moral proximity, and caring relationships.

Philosopher Arne Næss was discussed in Chapter 5, and we return to him again to take up other aspects of his work. He is perhaps most famously known for coining the term "Deep Ecology". In using this term, he was imploring everyone with environmental concerns to ask deeper questions. This is what philosophers do to try to understand fundamental aspects of an issue – the values buried deeply beneath the surface. They ask those pesky and persistent "why?" questions. Why do you think this way? What are your assumptions? Once some tentative answers are given, another round of questions is asked. Why do you hold these beliefs, assumptions, values…? Once exposed, these assumptions and values can be debated, evaluated, weighed, acted on, or replaced. Much of this book is related to processes like this, but is it enough?

"Is deep questioning sufficient?" was a question put directly to Næss.[1] As it turns out, for him, it was not. For Næss, it was important to get into a practical situation, to do something together. Out of these actions challenging perspectives can arise. Using his experience as a mountaineer, he was able to relate a famous story:

> I was climbing a little with a strong supporter of Hitler in 1935. I had some pieces of bread and I said: "This was baked by a Jewish girl. See if you can eat it anyhow." Then he admitted: "Well I do not mean that absolutely every Jewish person is a terrible so and so. There are exceptions." With reluctance he would then eat just a little of the bread.[2]

For Næss, it was the inescapable imperative to act – in this case to take some bread or not, which is also an action – that enables people to engage, and even confront, their deeply held values. For him, this may even enable a person to change the formulation of what they have seen as absolute truths. Educators sometimes call this transformative learning.

When asked about the subsequent holocaust, Næss responded, "when people saw that things were going very wrong they didn't stand up as much as they should have."[3] That is, there were not enough people prepared to share their bread. There are lessons for us in this observation. This first section of this chapter focuses on ethics as action.

The second section explores proximity. The idea of being close to something can be seen in many ways, and readers are encouraged to explore these possibilities. In this example, there is proximity, or a physical closeness, between the two climbers, and through the medium of the bread, the young Jewish woman. Nobody is abstract in this example, and the bread makes the girl real, too – it is the product of a real person.

The third and final section examines relationships and care. There are elements of Næss's story that exist in relationships. The two men have a relationship of sorts. They may not be close friends, but they are both climbers and they have entered into an outing together. The bread is also relational. It is the product of a young baker and is a source of sustenance for the climbers. It brings everyone together. And, once together, it challenges at least one in the party to evaluate his actions, his way of being in the world, and one of his truths.

---

[1]   Arne Næss and Bob Jickling, "Deep ecology and education," 48-62.
[2]   Ibid., 51.
[3]   Ibid., 52.

# Ethics as action

## Catch and release fishing[4]

One story about ethics "in and as" action, is taken from the *Regulations Summary for Fishing in the Yukon* in Northern Canada. These annual regulations outline rules for angling. The story is about the appropriateness of "catch and release fishing," that is using a rod, line, and hook to catch fish that are landed by the angler and then released back into the water. Given difficulties associated with regulating "catch and release fishing," a creative approach is taken in reporting the challenges inherent in this issue.

Rather than relying upon rules, these summaries tell the catch and release story from multiple perspectives. For First Nations peoples, it has often been important to tell their stories in their own way and, similarly, other anglers have stories that they want to tell too. Presenting these multiple stories alongside each other can provide a useful starting point for deliberations. For example, First Nations peoples often remind us that, "Fish are to eat, not to play with", or "The fish comes to you as a gift. It's offering its life to you. If you don't accept it, that's an insult. Sooner or later the fish will stop coming to you." Other anglers, supporting catch and release fishing, suggest that, "for me a fish is priceless too. I can't put a value on the peace of mind I get when I go fishing. I can't put a price on how important it is to me to be with my family: my son, my daughter, my wife, in the kinds of places where you find fish." These kinds of statements are supported by information about methods for careful release of fish, "live release ethics", and the "use everything – waste nothing" tradition of First Nations peoples.[5] This may seem irresolvable, but it does not have to be.

For years these contrasting stories have been presented together. Each time an angler flips through the regulation booklet he or she is invited to consider them. Put another way, the public is invited into a personal discussion about ethics and to consider how she or he will respond. The authors of the regulations have worked towards presenting the stories in a non-judgmental way – that is various perspectives are presented together in an open-ended fashion. For many, this opens new space for ethical deliberation and new possibilities for practice. Each year revisions to the previous booklet suggest some movement and creative interpretation, as one subsequent respondent said, "At the end of the day, we are all Yukoners. We share a common resource and the foundation of our thinking on both sides is respect." Some readers may critique this statement, but we do not

---

4    From Bob Jickling, "Ethics research in environmental education," 20-34.
5    These comments do not represent a systematic survey of the fishing regulations, though that would be a good project. Rather, we've sampled the regulations published between 1998 and 2004 by Yukon Environment.

think it should be considered an end, or a product, rather it is a snapshot of an individual engaged in a process that is ongoing.

There is however one final step to this story. The angler must decide what to do when landing the fish – when the person and the fish come into physical contact. Will the angler keep the fish for food or release it? How will the ideas presented in regulations – or drawn from other sources – affect his or her judgment and actions? How will knowledge about size, breeding potential, or survival rates for released fish affect her or his decisions? How will actions – in releasing or keeping the fish – affect his or her personal ethics?[6]

## The ant and me

Saransh Sugandh shared a different kind of ethics as action story.[7] He is an Indian environmental educator who attended an ethics workshop on a draft version of this book. He tells about an intervention by his grandmother while he was in the act of squashing ants with his flip-flops:

> As a child, I used to love killing insects; the smaller and more non-threatening the better. Ants were a particular favourite. We used to live as a joint family in a three-floored house with the space on one of the mezzanines being used as a small granary. It would store rice and wheat that would come from my father's village. One day while playing in the staircase, I noticed a trail of ants leading from the vent of the granary, along the staircase and into a small hole in the wall.
>
> I took out my flip-flops and went smacking the ants dead. I imagined what a catastrophe it must be for the ant colony – their evening news full of horror as tens of them were attacked by aliens. As my younger brother watched, trying to comprehend if it was fun or not, as I went about the alien attack.
>
> That's when I heard my Grandmother call out, "Sugandh, what are you doing?"
>
> "Killing the ants, they are taking the rice from the granary," came the very sure reply.
>
> "Stop killing them! Rama[8] loves them. The ants are his *vahana* (in modern parlance, a vehicle)."
>
> "Hain!" I stopped after a very sure splat on the wall.
>
> "*Matlab?* (meaning?)" I turned around.
>
> "*Matlab ki,* (means) the god Rama uses them as his *vahana*, so you must not kill them. The gods will be angry."

---

6   This issue has also been explored in a more "traditional" way in the journal *Environmental Ethics*. See, for example, A. Dionys de Leeuw, "Contemplating the Interests of Fish," 373-390.

7   Saransh Sugandh is a filmmaker, environmental educator based in Delhi. He currently runs his own firm called In Vaarta Communications.

8   Rama is a Hindu God, hero of the mythological epic *Ramayana* 11.

She stood there at the landing of the staircase, wearing her usual white sari. She was a widow and had been for a long time. She was deeply religious and I knew she meant what she said. I believed in her stories like I assume most six year olds believe their grandparents.

The alien had stopped the attack. As I saw the black ants scurrying away, I saw their existence with renewed meaning. My gods had a place for them as well. They were not alien.

As I learnt later, it was not just Rama and the ants; the elephant-headed god Ganesha rode the mouse; Saraswati, the goddess of learning and knowledge had swans; The goddess of wealth, Lakshmi known for her slippery feet – rode the owl; Garuda or the eagle-like bird had Vishnu and even snakes had a protector in Lord Shiva.

Then how could we kill anything? It was difficult to see any life form as alien anymore. Though my family was a meat-eating one, I turned vegetarian soon after.

*(I started eating meat again, off and on, twenty years later to be able to share my then partner's food. To her, it was important to be able to share the food she had prepared and so I agreed.)*

Like the Yukon fishing example, this is very much a story about ethics arising through action. Saransh's grandmother's intervention was not delivered as an abstract lesson or a bedtime story. Rather, it was an admonishment while Saransh was still gripping his ant-squashing flip-flop. With a fresh perspective about ants still ringing in his ears he had to decide whether to put the footwear down or use it again in his ongoing binge of ant-splattering. He put the shoe down in what was, ultimately, a life-changing ethical action.

## LEARNING ACTIVITY 8.1: EXPLORING POSSIBILITIES FOR ETHICS AS ACTION

1. Consider the examples given here – catch and release fishing and ant killing. Can you think of times when you have learned something about your own values when you had to choose between different ways of handling a situation? Think about your daily activities. Are you practicing ethics in action? Perhaps you can share some examples?

2. Who is a role model in your community? Identify examples of others who are re-personalising ethics in their actions – through their work or their lifestyles. Or identify examples of those who take responsibility for the other (human or more-than-human) without expecting reciprocity.

## EXTENSION ACTIVITY

3. Consider what kind of "ethics as action" activities can you undertake in your own backyard?

4. Do some lifecycle research, such as tracing a given product from initial production through to final usage,[9] to find out how you could change your daily lifestyle choices. Try researching one of the following items, or something else you use/buy a lot: a hamburger; a cup of coffee; a T-shirt; a pair of jeans; or a newspaper. Write a story with the title "The secret life of ...," and tell the story of the different ethical choices you are making when you consume/use the product in your story. What alternatives are there? Start a display in your classroom, tearoom, or home to inspire others to do the same!

## Moral proximity and ethics: An attentive struggle in a modern world of ambivalence

In this section we will expand on the idea of ethics as action, but frame this expansion through ideas about moral proximity. Sometimes, the place of moral impulse, proximity, and an expanding ethic-of-care in driving ethics-as-action can be overlooked in learning situations. This is especially true where ethical questions about the common good and future sustainability seem complex, stubbornly difficult, and perhaps even intractable.

As many examples in earlier chapters of this book show, ethical deliberations on social-ecological matters of concern can, themselves, produce compelling re-imagining. Environmental educators sometimes respond to animal welfare concerns through deliberations about the cruelty and health risks associated with modern factory farming. Outcomes often include a resolve not to buy foods produced in these morally indefensible ways. The more militant amongst us have been inclined to expose learners to the harsh realities of factory farming. But, as discussed in Chapter 6, philosopher Zygmunt Bauman suggests that this is not enough.[10]

For Bauman, learning processes limited to ethical deliberation do not include the physical proximity and practical enactment necessary to trigger moral impulses. That is, the moral impulse to take action might often not be realised, especially in the overwhelming face of ways of knowing, being, and doing things in the modern world. Bauman's commitment to being attentive and disposed – to simply being for the Other – opens the way to reflexive contemplation that is combined with serious moral intent to enact ethical practice.

---

[9] For ecologically led lifecycle resources, check out William McDonough and Michael Braungart's Cradle 2 Cradle institute, www.c2ccertified.org/about, as well as Annie Leonard's Story of Stuff Project, www.storyofstuff.org

[10] Zygmunt Bauman, *Postmodern ethics*.

This section concerns enabling the moral imperative through concern for the Other. Here, extending this circle of concern goes beyond just humans, to include the other-than-human, and wider world systems. We use two case stories that explore the moral impulse and strengthening imperatives that can emerge with practical activity – hence proximity – to develop ethics-led learning and learning-led change in response to challenges and contradictions embedded in everyday environmental practices. In both cases, a sustained period of attentive activity made possible an open-ended process of working together that enriched engagement, and even re-enchantment, in the world.

The practical cases of ethics-led learning examined in the case stories open some interesting dimensions of proximity that arise from physically taking action because of concern for the Other.

The first case story explores a household's engagement with the problem of factory-farmed eggs. The second case story illustrates the challenge of extending concern to wider systems, in this case waste and matter cycling. It is situated in a group setting, and, in the end, culminates with the transgressive challenge of preparing a nutritious, zero-waste meal.

### Case Story 1 – Factory eggs: An expanding moral imperative within a household story

Bauman has described how modernity has shaped a rule-governed and distant world where people are, for example, far removed from the production of consumer goods. So removed, they can be ambivalent towards, and even blind to, the dark side of modern practices. Our interest here is in how taken-for-granted, everyday practices can be opened in more proximal ways. That is, we are interested in bringing the realities of everyday practices closer to the lives of learners in ways that they can be better engaged in personal choices and actions.

In this case study, the everyday practice of buying eggs, which most of us get from a supermarket and consume in the normal course of our daily lives, is opened to practical scrutiny. The following narrative, by Rob O'Donoghue, sketches a household account of bringing chicken rearing into close proximity:

> After deliberation on the ethics of modern chicken farming, my wife, Carmen, began to take great care to buy farm fresh eggs where the chickens have not been subjected to the sufferings related to being turned into caged egg-producing machines, a condition that she finds morally repulsive. As more people come to feel this way about the production of what they buy to eat, those farming and marketing eggs have adjusted their ways of doing things and of representing the product so that the story of how the food comes to the plate is more palatable.

The difficulty of assuring the quality of the eggs we consume and the treatment of the chickens has led us to try keeping and rearing our own egg-producing chickens. I thus made a low-cost pen following the same "chicken tractor" design that I had used some years earlier. This earlier experiment with chicken keeping had not been a pleasant or successful experience. The first time we made a pen for the garden, the dog made merry with the chickens in a killing frenzy.

Figure 8.1    Return of the chicken tractor
(Photo by author, Rob O'Donoghue)

It took the old dog passing away and our children leaving home for us to try raising chickens again. This second try was equally traumatic as one of the chickens turned out to be a rooster who harassed our new neighbours who were navigating the sleeplessness of the arrival of their new baby. We were thus left with one hen who took to us as company and produced a reliable stream of eggs – before running out of steam and wanting to sit on eggs to produce the next generation.

We did not want to subject our little red hen to the experience of sitting in false hope so we searched for a source of fertile eggs. Our own hopes were in being able to raise the next generation of egg producers in and for our family. At the organic-foods store we learned that although the eggs were humanely produced, there was no cockerel involved, as most consumers of organic eggs do not like to have a fertile spot in their sunny-side-up delight. Luckily, contact with a local farmer's wife had her kindly supply us with half a dozen fertile eggs for our broody hen.

Presenting these fertile eggs to our hen, as she sat in sulky silence on an empty nest, was a delight. She puffed up and clucked with satisfaction as I presented her with each egg in turn. Her head disappeared entirely as she took each egg and tucked it under herself, seemingly developing an inflated sense of self-important satisfaction as the number mounted. The experience also gave me a great deal of satisfaction. I shared the news with Carmen in an exchange of text messages while she was away visiting our grandchildren. However, I did not share what happened until after she returned.

The next morning our little red hen was still puffed up and self-important but she had managed to kick all six of the eggs out of the nest, and four of them had been cracked with the fall. I thus had a large omelette for breakfast that morning and then modified the nest so that the inexperienced shuffling of a new chicken mother-to-be did not eject any more eggs.

It became a concern to me that our little red hen was a novice without the experience of looking after a clutch of eggs, or herself. As it turns out, the expected confinement, from sitting to hatching, is 21 days. This information was discovered by our grandchildren and Carmen using the i-Pad and then emailed to me.

In the meantime, our hen had already been sitting in stony silence for a week before we got her the fertilized eggs. I did not want her to lose condition so I took the radical step of lifting her off the nest every day so that I was assured she was getting enough food and water. She took exception to this and would sit indignantly for a few minutes before stiffly getting up, venting her anger with a series of squawks, and flapping off across the lawn to drink deeply from the cats' water bowl. She would then feed on some grain that I threw down for her and amble off into the vegetable beds where she would scuffle for bugs before having a dust bath and returning to the nest.

Three weeks later, we were witness to the arrival of a lone chick who looked surprisingly like her surrogate mother. Not fluffy and yellow as on TV but a motley yellow grey with a cheeky face and little tufts of feathers on her legs. She was to be an only child as the other egg had seemingly been damaged in the earlier fall, and a chick had not developed inside.

Figure 8.2    Our first chick (Photo by author, Rob O'Donoghue)

Once the problem of the second egg had been resolved, we could give attention to our now sickly and weak-looking chick that did not seem as if she would survive. Also, she was infested with lice. The fading eyesight of age had not allowed us to immediately notice that flea-like bird lice had proliferated in the nest and now covered the chicken, the chick, and the two of us.

Once again the internet enlightened us; wild birds can bring in lice. The local chemist came to the rescue supplying a dusting powder that seemed to resolve the problem for the mother and baby. For a further day, the chick continued to stagger about and fall over before starting to feed and gain strength. It was delightful to watch the young and inexperienced mother hen drop bugs that she scratched up from some compost that we put in the pen, and drop these before the bewildered young chick.

We, however, were still scratching as we watched these delights. So, we had to get medicated shampoo for ourselves, and spray our bed, to finally rid ourselves of the lice and fleas that accompanied the experience.

Now we still have a problem that there is a 50-50 chance that the chick might be a boy, and we may have to resolve the problem of a cockerel once again. So it is back to the Internet for information on the sexing of chicks. Few people raise chickens in urban gardens in our area, and we have little in the way of life experience to draw on, so the Internet has become the source of our explanatory wisdom.

In reflecting upon this experience, O'Donoghue later remarked:
In this case, a questioning perspective on factory farming shaped the moral imperative for our household. Within this imperative, and through our chicken rearing, we were able to clarify an ethical

disposition that guided our way of being in the modern world of mass-marketed eggs. Educationally we noted how modern attitudes of relatively blind consumption were brought into question. This questioning opened the way to reflexive change achieved through a deliberative and experiential resolution to the matter of our concern.

In some ways, this case story is like earlier examples in this chapter: Næss's example of sharing bread with a Nazi sympathiser, or a decision about participating in catch-and-release fishing. In all instances, the participants are close to the questions at hand; there is a moral proximity that must be negotiated. There is literally a piece of bread, a fish, or an egg in a person's hand, forcing a decision to be made.

This egg-rearing case story is particularly interesting in describing a longer-term engagement with the ethical conundrum at a household level. This egg-in-hand experience demanded deliberative reflection on the problems of health and cruelty in factory farming. For this family, it produced a moral imperative on behalf of the Other – in this case chickens. This imperative was initially directed towards the alternative of buying only free-range eggs. However, the reliability of eggs marketed as "free range" was also brought into question.

A further step on behalf of the Other, keeping chickens at home, produced more experiences of ambiguity, complexity, and uncertainty that were difficult to navigate against an imagined "pastoral ideal" that did not match with reality. The initial deliberations on the merits of free-range eggs, with all the associated complexities of defining exactly what this means, took their household to a position of being for the Other that ultimately extended to an ideal of trying to keep their own chickens.

The pastoral ideal was an elusive feature in the process, but this was modified in successive re-enchantments that came with the, at times, harsh adjustments required by the realities of raising chickens for eggs. By the end of the reported narrative, the household had arrived at the cusp of a more satisfying way of being – more ethically attuned to expanded limits of reason. Through this example of ethics-in-action, Rob and Carmen transgressed much of the ambivalence, and many of the contradictions, of modern factory farming. Though they are now engaged in a changed way of doing things, many other contradictions still demand their attention. For example:

- How much more is each egg costing us?
- How do we ensure a regular supply?
- Can we reduce the cost by feeding the chickens by using vegetable scraps?
- How do we integrate the use of the chicken tractor with our vegetable gardening?
- Is it the carrot peelings that are making the yokes that wonderful deep orange colour?

These reflections scope contours of learning to change everyday practices within a household. The process began with discomfort arising within an informed pattern of practice that was at odds with a disposition to care for the Other. In turn, this gave rise to a moral imperative to do what Rob and Carmen felt was the right thing – that is their ethical standpoint. Attempting to bring their practices in line with their ethical standpoint, brought both new proximities of re-enchantment and challenging realities that contradicted imagined ideals. Still, many of the problems were resolved in an unfolding story that, in turn, brought further challenges. Their continuing ethic of attentive struggle is in the company of the chickens they now keep in their garden.

## LEARNING ACTIVITY 8.2: CHICKEN AND EGG

1. Consider the case story presented above. What is the moral driver of the exploratory changes in practice? What conflicting interests or tough choices are reported in the story? What was behind the desire to do things differently so that the problem of cruelty might be resolved?

2. Are there ways that this problem could be responded to in your context? What are some of the new challenges that might arise if you tried to take this idea into action? Why is it important to you to come up with a better way of doing things in the world?

3. What environment and sustainability thinking is behind the need to work out a better way of doing things? Why do this? What are some of the challenges that you might expect if you tried to act on your ideas? What would make taking up the challenge of change worthwhile? Why is it worth trying?

### Case Story 2 – A zero-waste engagement with the modern problem of recycling

This case story builds on the social complexity of stories such as the one presented above. Stories of that type play out at an individual or household scale. However, what additional complexities will arise within a larger social setting? For example, how can an educator guide an action-oriented project at the scale of a whole class? In this case, "waste" is explored with the inevitable vagaries that can arise within a community.

The problem of waste has been a pervasive topic in environmental education. Here most school children have been taught to "reduce, reuse and recycle" as a mantra for a modern lifestyle and a response to problems of ethical waste management. In most eco-school learning programmes, children are challenged not only to creatively re-use waste in practical articles and artworks, but to become avid collectors and separators of waste for recycling. The emphasis on recycling, as a solution to waste, is often driven by the waste economy of the plastics, "tin", and paper recycling industries that claim to return the materials back into material cycling in a way that can support the health and convenience lifestyles of modernity.

Then along came the pressing issue of climate change to "burst the bubble" and prompt a closer scrutiny of waste streams. Climate change "lifted the curtain" on the carbon in activities like recycling. A closer inspection typically reveals how significant amounts of energy are used, and toxic pollutants are produced in separating and reusing waste. Also, in practice, insignificant amounts of the waste stream branch back into a productive economy. Sadly, to increase the amount of material that could be reused would produce an inordinately complex problem of increasing climate change gas emissions and toxic pollutants. Today most separated waste comes to reside in landfills, and more and more waste accumulates in the global gyro. Despite this paradox, children are still taught the same mantra and efforts to surface contradictions have not been easy.

The following narrative describes Rob O'Donoghue's engagement with these contradictions with his class at Rhodes University:

> When I was asked to encourage "first years" at university to do the right thing and separate their waste for recycling, I found that they had already been effectively taught this throughout their schooling. I thus attempted to shift the focus from recycling to reducing waste so as to point to the pollution and carbon emissions produced by an untenable waste trade. This was greeted with shocked indignation, and a protracted deliberation followed. Little came of this other than a change in the recycling mantra to "refuse, reduce, reuse and recycle."
>
> My next attempt played on this new mantra when I asked, "What would we have to do to be able to avoid recycling everything but organic waste? And why might this be a good thing?" Further shocked indignation and debate followed and I was marked as a somewhat radical environmentalist. Trying to live up to this brand, but not being keen on campaigning against recycling, I contemplated the positive challenge of trying to produce a zero-waste meal as a way of engaging the students in the waste recycling paradox.
>
> Contemplating and deliberating this challenge proved to be all but impossible; recycling as an ethical practice was deeply entrenched. At best the deliberation concluded with the impossible scenario of having to end up only eating what you can grow and scavenge. The circulating conversation had us decide to take up the challenge of preparing a zero-waste meal.
>
> To take up this challenge, we had to contemplate a healthy but currently wasteful meal like pizza, salad, and fruit juice. The choice of a pizza meal gave rise to a heated debate between an omnivorous and a vegetarian diet, and which was best for the environment. This question was sensibly "set

aside for now" and must remain beyond the scope of this chapter, but the decision was taken that a wholesome meal should include carbohydrates, proteins, fruit and vegetables, as well as cater for vegetarian preferences.

Good research needs a contextual baseline, so we elected to buy and share healthy fast food and supermarket meals of pizza and salad with fruit juice. Everyone enjoyed the food, but were shocked by how much waste our tastes and purchasing practices had produced.

Figure 8.3  A shocking amount of waste (Photo by author, Rob O'Donoghue)

As we enjoyed the food together, we deliberated about what we would have to do to prepare the same meal with zero-waste. This turned out to be a relatively simple matter of asking **what we would have to do differently** to have not ended up with:

- **Two pizza boxes.** Done differently: Bring your own container or make the dough and pizzas yourself using fresh ingredients.
- **Plastic knives and forks.** Done differently: Eat with hands.
- **Paper napkins.** Done differently: Wash hands before and after eating.
- **Fruit juice box.** Done differently: Squeeze fresh juice.
- **Vegetable trays and plastic wrappings.** Done differently: Buy without packaging.
- **An empty bean tin.** Done differently: Buy at deli using own container.
- **Payment receipts.** Done differently: Pay cash and buy from small shops and street vendors.

We managed to come up with a whole range of hypothetical things that could be done differently "to absent" each of the waste items but probing questions always surfaced complexities where waste would creep in. For example, we hypothetically "absented" everything until it came to the receipts and then realized that most shop purchases would be accompanied by unwanted paper.

The next question became, who could actually try to do that for one meal at home during the following week. As all the participants were students living in residence, they agreed to explore ways of shopping to reduce impact on the waste stream. I, too, accepted the challenge of the zero-waste meal. With some rapid sleight of hand, I had managed to renegotiate the challenge of preparing an as-close-to-zero-waste meal as possible. I also managed to negotiate the concession that composted waste that went back into the garden or an item that would genuinely go into continued practical use did not count as waste.

When I got home from work that evening, Carmen said, "Rob, what have you got us into this time!"

It was relatively easy to find alternative shopping sites and to negotiate no packaging ways of sourcing the ingredients needed for making pizza to the standard that Bob Jickling demonstrated during his stay with us on one of his trips to the Environmental Learning Research Centre at Rhodes University. Small changes like bringing my own containers was relatively simple in theory but was not the norm at supermarkets where most things are pre-packaged. But, at Lungi and Ingram's Farm Stall I could buy fresh ingredients without packaging. One of the more difficult issues was the complexity that all of the staple ingredients like flour, yeast and salt that one takes for granted at home, had all come in packaging. I thus decided to try to source these and all other ingredients in novel ways for the one meal challenge.

All of the seemingly practical ways we had thought out for enacting a zero-waste pizza with a bean salad and fruit juice were not easy in our busy world of sedimented daily routines. Many seemingly simple solutions that I came up with had a knock-on effect that produced subtle waste complexities that there did not seem to be any way of detouring, like dried yeast coming in a sealed foil packet. There seemed to be no easy zero waste options but a heartening thought was that each option explored had considerably less waste than our bench-marking meal from the pizza shop and supermarket. And the creative exploratory process had me making many new friends who were happy to help, like the local sourdough baker who sold me a lump of wonderful wild yeast dough and taught me to make my own with home ground wheat. The local green grocer supplied most things fresh and unpackaged as a matter of course and the gourmet deli was happy to pop the cheeses into my container. With a few lucky breaks and creative innovations like this, the best options were:

- Buy salad greens, carrots, celery, beans, tomatoes and juicing fruit from Lungi.
- Soak and cook dried sugar beans from the garden the day before.
- Buy cheese using own container from Virginia at Fusion Deli
- Buy a reusable glass jar of anchovy fillets and the same of capers from the Pick & Pay Deli. (The contents keep for months in the fridge and we will make jams in the small jars that are ideal for our own use or for giving as gifts for friends).
- Source sourdough from Jacque, the friendly artisan baker, or grind the wheat myself (sourced in a cloth sack from Lesotho) and make my own, 4-5 days before.
- Cook everything on a pizza stone over a charcoal stove.

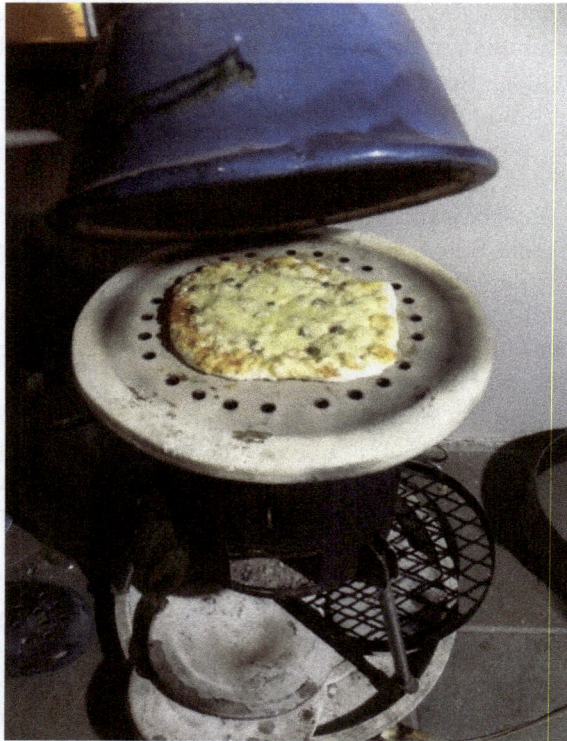

Figure 8.4 Charcoal pizza oven
(Photo by author, Rob O'Donoghue)

The latter gave the whole enterprise a rustic charm and a close to the earth feel so, on the designated day, I set off from work on my bike to source all of the ingredients on my way home. What a delight that was, on a cooling autumn evening, stopping here and there to make the purchases and to explain the reasons for the odd requests like not to have a printed till slip. This was something that was not always possible, but with a little creative manipulation I was able to satisfy the zero waste challenge by popping the paper into my prolific compost heap that will eat up anything that is close to organic.

The result was very satisfying and on the menu we had:

- Crusty sourdough pizza with a 3 cheese topping on anchovy tomato puree or basil pesto;
- fresh green salad with celery, carrot and tomato along with sautéed green and kidney beans; and
- freshly squeezed carrot and orange juice.

The carrot pulp and vegetable waste went to the chickens who were delighted with the windfall, scratching and pecking at it vigorously.

Figure 8.5  Vegetable waste for chickens
(Photo by author, Rob O'Donoghue)

As we sat enjoying our pizza, bean salad and fresh orange juice, Carmen, as usual, asked an incisive question to conclude an exploratory week that culminated in one low-to-zero-waste meal. Would it not have been a lot easier to take our own containers to collect pizza and salad from the pizza joint? I had to concede that the answer was a resounding YES, but it wouldn't have been half as much fun and I wouldn't have ended up with a mini charcoal pizza oven for zero-waste meals.

Figure 8.6   Ingredients for our low-to-zero-waste meal
(Photo by author, Rob O'Donoghue)

This case story is presented with a lot of good humour, but also with serious engagement. In contrasting this story with the first, this one demonstrates that while it can be one thing to make changes in household practices, it is quite another to challenge a group to rethink well acculturated environmental practices. Both stories clearly show that enacting ethical ideas and practices are always complicated, in practice, and require on-going attentiveness.

## LEARNING ACTIVITY 8.3: ZERO WASTE ETHICS

1. Carefully consider this second case story and identify the emerging contradictions, and the moral driver of the exploratory change practices reported in the story. What was behind the desire to do things differently so that the problem of wastefulness might be resolved (absented)?

2. Can you think of different ways to respond to the problem in your context? What are some of the new challenges that might arise if you tried to take your ideas into action and why would it be important to you to come up with a better way of doing things in the world?

3. Consider one the following challenges and discuss the questions that follow.

   Choose a challenge:
   - How would I go about **cooking a meal** that is healthier, more fun to produce and tastier than my favourite fast-food treat?
   - How would I go about **choosing toiletries and cosmetics** that are produced locally, are not tested on animals and do not have a polluting impact on the environment?
   - How would we go about **growing the food for a meal** to feed a family?

The two case stories reflect an engagement with moral questions in a modern age of consumerism. They raise ethical questions that need careful deliberations that can then be enacted, with the complexities of real people and places. Through exploratory action, these case stories are of value where they become "stories that do work". The questions asked here may enable you to work with the stories to explore your own questions that arise around moral dilemmas and ethical expansion in a reflexive context of proximity and social innovation.

## EXTENSION ACTIVITY

4. Try out one of the activities that you have discussed in Activity 3 (above) and keep a diary so that you have a story to share. Remember, we cannot expect to be fully successful the first-time round, but every effort at change usually changes things for the better in some small way. It also usually makes clear that the effort has been worthwhile and what challenges to change still lay ahead.

## Relationships, identification, and care

There are other examples of relationships in the world that are quite different from those given in the previous two sections of this chapter, however they are profound. For example, Norma Kassi, a Vuntut Gwitchin First Nation member, talks of a time when she was six years old and travelling on the land with her grandfather. It was a time where there were lots of wolves around. She said:

> All we knew is that there were lots of them. At that time too there was still no caribou. Anyways those wolves stayed in the bushes about four days. My grandfather sat outside the whole time and finally on the fourth day early in the morning he walked over there and I thought I would never see him again. He walked into the bush and talked to them. Pretty soon all of the wolves were moving out. They started going down onto the [frozen] lake. We could hear the willows and everything breaking. It was an experience that I will never forget. Not once did any of those wolves try to hurt us. They never came near our camp; they just stayed over there. My grandpa just sat there and with his own quiet powers was able to communicate with them.
>
> Two days later my grandfather said that there was not going to be any caribou here this spring, we have got to move. He knew that the wolves were following the herd and there was not going to be caribou here. So we moved. We only got a couple of caribou that spring, and that was it.[11]

Kassi's story points to the possibility of a kind of intimate relationship with other beings. Yet, she says it is hard to understand when we try to explain our experiences. What then would it mean, for example, to be in a relationship so profoundly intimate that a kind of interspecies communication can take place?

Kassi also talked about the difficulties in talking about her culture's beliefs and spiritual way of life. For her, it must be lived, experienced, and seen. However, to experience a different way of being in the world goes beyond what most people experience within "Western" or "European" styled educational experiences. While her story is vastly different from the earlier one about Næss sharing some bread, or Rob O'Donoghue's chicken rearing, one common element is the learning that arises from being immersed in and confronting life's experiences.

It is true that understanding Kassi's cultural and spiritual way of life is infinitely more complex than we can talk about here, and, ultimately, that is her story to tell. However, in her choosing to tell this story, we believe that she wanted us to know that there are other ways of living and being in the world, different from the ones that most readers of this book see most often. These ways of being involve intimate relationships with the

---

[11] Norma Kassi, "Science, ethics and wildlife management," 212-216.

more-than-human world in ways that seem almost unimaginable in most educational contexts of the 21st century.

There are, however, examples of people that have caught glimpses of radically different ways of being in the world. When they have, their experiences have stayed with them for a lifetime and been a touchstone for their writing, professional practice, and lifestyles. Some well-known examples are discussed in the text below. For the first, we return to the work of Arne Næss for a Norwegian example and the second is from the American, Aldo Leopold.

Additional examples are provided in the handout page. One of these is a famous description from a novel by Albert Camus. The fourth story is not famous, but just as important. It is from a South African blogger.

\*\*\*

When Arne Næss pursued his own deep questioning process, he found the basis for his inclination to act came as identification with parts of the world – particularly parts of nature. In this identification with the larger world, he found joy, love, and the inspiration to act. Næss did not find the impulse to act in abstract principles and duties. Rather it arose from a loving realisation that he was not an individual self, but part of a greater whole, a "Self" much larger than his egotistical "self." In his recognition of this greater Self, Næss realised humans could not exist without the other in beautiful, loving relationships.

In describing this realisation, Næss often tells of an experience early in his life:

> What would be a paradigm situation of identification? It is a situation in which identification elicits intense empathy. My standard example has to do with a non-human being I met 40 years ago. I looked through an old-fashioned microscope at the dramatic meeting of two drops of different chemicals. At that moment, a flea jumped from a lemming that was strolling along the table and landed in the middle of the acid chemicals. To save it was impossible. It took many minutes for the flea to die. Its movements were dreadfully expressive. Naturally, what I felt was a painful sense of compassion and empathy, but the empathy was *not* basic; rather, it was a process of identification: that "I saw myself in the flea." If I had been *alienated* from the flea, not seeing intuitively anything even resembling myself, the death struggle would have left me feeling indifferent. So there must be identification in order for there to be compassion and, among human beings, solidarity.[12]

---

[12]  Arne Næss, "Self realization: An ecological approach," 22.

It is interesting to note that while compassion and empathy are described as part of this experience, the flea is not an external other to feel sorry for. Here, Næss describes a deep relationship with the flea, one that does not separate the human from the more-than-human, but brings them together into a single identity. A capacity for such an integrated identification would indeed reflect a profoundly different way of being in the world – a different ontology. Perhaps, such ways of being in the world would even resonate with the ways of being in the world implicit in the story related by Norma Kassi in the text above?

***

American conservationist, Aldo Leopold wrote of a similarly life changing experience. As a young man, he was involved in the killing of a wolf. He said,

> Only the mountain has lived long enough to listen objectively to the howl of a wolf … My own conviction on this score dates from the day I saw a wolf die. We were eating lunch on the high rim rock, at the foot of which a turbulent river elbowed its way. We saw what we thought was a doe fording the torrent, her breast awash in white water. When she climbed the bank toward us, and shook out her tail, we realized our error. It was a wolf.
>
> In those days we had never heard of passing up the chance to kill a wolf. In a second we were pumping lead into the pack, but with more excitement than accuracy. When our rifles were empty the old wolf was down and a pup was dragging a leg into impassable slide rocks.
>
> We reached the old wolf in time to watch a fierce green fire dying in her eyes. I realized then, and have known ever since, that there was something new to me in those eyes – something known only to her and to the mountain. I was young then and full of trigger-itch; I thought that because fewer wolves meant more deer, that no wolves would mean hunters' paradise. But after seeing the green fire die, I sensed that neither the wolf nor the mountain agreed with such a view.[13]

Watching this wolf die was a life changing experience, much like Næss's experience with the flea. The feeling it engenders seems to infuse the balance of his life's work. Here again, proximity and first-hand experience are at the core of the transformation. Leopold describes the relationship between ethics and this transformation in the following way: "We can be ethical only in relation to something we can see, feel, understand, love, or

---

[13] Aldo Leopold, A *Sand County Almanac*, 138-139.

otherwise have faith in".[14] We need to pay attention to this. Amongst other things, this might also be a harbinger of ethics of care.

It is also interesting to speculate about how hard this experience was to put into words. At first glance, many people can indeed empathise with the evocative image of green fire dying in the eyes of the wolf. But what is he trying to say, exactly, when he speaks about an animate mountain? Here, too, Leopold seems to be pointing to possibilities for intimate relationships with the land that go beyond our linguistic capabilities to describe.

---

[14]  Ibid., 251.

# HANDOUT:
# DIFFERENT WAYS OF BEING IN THE WORLD

The following excerpt is from the novel, *The Plague*, by Albert Camus. It is situated in an Algerian town that is suffering from an outbreak from bubonic plague. It was written just after World War II as an allegory of the German occupation of France[15]. Interestingly, it could be read today as an allegory for contemporary engagement with climate change issues. In this example, a character in the story was taken to work for a day by his father. In this case the boy's father was a prosecuting counsel, and on this day his father was to present an argument against an accused man. The excerpt begins at the point where the boy first noticed the accused:

> But, I suddenly became aware of him, though up to then I had only thought of him in the convenient category of the "accused." I cannot say that I forgot about my father, but there was something in the pit of my stomach which distracted me from concentrating on anything except the man in the dock. I heard practically nothing. I felt that they wanted to kill this living man and an instinct as powerful as a tidal wave swept me to his side with a sort of blind obstinacy. I only properly woke up when my father began his speech for the prosecution.
>
> Transformed by his red robe, he was neither good-natured nor affectionate and his mouth was full of sonorous phrases which leapt from it constantly like serpents. I realized then that he was asking for the death of this man in the name of society and that he was even demanding that his head be cut off. In truth, all he said was "This head must fall." But the difference was not so great in the end. In fact it came to the same thing, since he got the man's head. The only thing was that he did not do the job himself. I who subsequently followed the matter through to its conclusion, felt a far more terrifying intimacy with that unfortunate man than my father ever could. And yet he had to be present, according to custom, at what are euphemistically called the last moments, and which one should by rights call the most shameful of murders.[16]

* * *

The following excerpt was found on a South African's blog site. It describes an experience, and subsequent reflections, that occurred during the course of a day working with an anti-poaching NGO:

> I grew up in very traditional ways and slaughtering of cows, goats, chickens, and many other animals never really moved me emotionally because that practice has become part of who I am. My heart sank today when I had to watch a poached Rhino being treated face to face. I saw that beast lying there looking very unnatural because its horn has been removed in the most inhumane way. Poaching is a brutal exercise! What is interesting for me though is that those are the exact emotions I feel every time I go to eThembeni, Zolani, Phaphamani, Hoogenoeg, Vergenoeg and many other marginalised communities. Think about going to a bush every time you need a toilet. In a civilised society this is humiliating and it strips one's dignity off, it is brutal in many ways. There are many other things but I'll stop there. Maybe I feel like this because I experienced it first-hand and I still see many people, in fact the majority of South Africans, living in this way. These are the same people who are used by the multi-nationals and high profile politicians to perpetuate this act. This is the majority that can't relate with these animals because they don't own them and never interact with them. After all these people were removed from their homes (which were food producing farms) and the farms were changed to game reserves. Poverty, greed, corruption, capitalism are all responsible for all the injustices in the world.
>
> The injustice of Rhino poaching is but part of a bigger problem. The problem is systematic and structural. My heart is bleeding for the injustices done to the animals, environment and the people of the world. Ours should be a revolution for respect of humans, animals and plant life alike.[17]

---

[15]    Tony Judt, "A hero for our times."
[16]    Albert Camus, *The Plague*, 191.
[17]    Reproduced with permission from South African blogger, Xola Mali. The original blog post no longer appears available on the internet.

## LEARNING ACTIVITY 8.4: RHINO RELATIONS

1. Examine the learning that took place in these examples. What stories of transformation learning are familiar in your own region? In your own culture?

2. Think again about the learning that took place in these examples. Could these lessons be taught? Could they realistically be presented as learning outcomes? How are they different from the experiences in traditional educational settings?

3. What kinds of educational activities could you create to give a "taste" for the experiences described in these stories? What kinds of educational experiments could you try? What are some of the barriers that you might expect to face? But, what kinds of openings can you find to try out these experiments?

## EXTENSION ACTIVITY

4. Consider the experiences described in the handout pages. The excerpt from Camus's North African novel is famous. The South African blogger's story is less famous, but just as important.

   The Camus excerpt is interesting, in part, because it is from a work of fiction. Sometimes, fiction can resonate deeply with real life, as deeply as non-fiction. It can sometimes tell a story with more nuance than any theoretical description. Does the experience of the prosecutor's son resonate with you? If so, in what ways? In small groups, discuss your response to this excerpt. Can you find common experiences that link you to this description? What other examples of evocative descriptions of transformational experiences can you find in literature? Internationally, and in literature from your own region?

   The blogger tells of a profound experience with a rhinoceros that had its horn removed by poachers, but also how this experience was situated within larger systemic issues. This excerpt points to the possibility of something deeply transformational arising from this experience – an opening. Perhaps we can only really know if something is transformational when looking back and tracing one's actions over time. Can you find similar stories in your community that are potentially transformative? In small groups, share your own experiences. What experiences would you say have been transformational? For yourself? For others in your group?

### A few thoughts about pedagogy and transformational learning

This chapter has been about educational experiences that arise with fully embodied experiences in the real world. That is, this is about the importance of educational opportunities that are not generally amenable to classroom learning, lectures, or books. That is not to say that reading and classroom discussion cannot play a role in setting up these kinds of experiences, or reflection on them afterwards. However, at root, they all describe direct encounters with the world.

Another theme runs through these examples. They point to things that cannot be controlled, predicted, and probably not evaluated. For example, how can we possibly know if an educational experience has been transformational without allowing time, probably years, to gauge the impact of the experience on a person's life? Still, it seems to be educationally useful to provide opportunities, or openings to alternative ways of being in the world, as well as opportunities for direct engagement.

Chapter 5 provides examples of some approaches that could support experiential learning. They include bird watching, journal making, and other forms of environmental art and field trips. Earlier in this chapter, other projects were designed to bring moral proximity into the realm of learners' experiences.

In another approach, Bob Jickling built learning around action projects in some of his courses at Lakehead University. Here, learners were first asked how they might contribute to their community – how they might make a difference socially and environmentally. Second, they were instructed to research organisations and/or sites where they could enact their contribution by providing a strong social-ecologically-based service over the course of a semester.

In class, debriefing outcomes from these experiences varied. Most learners seemed to have gained some useful knowledge and human insight. Occasionally, someone would relate a story of their experience so powerful that the whole class would be moved – sometimes to tears. There are several things to consider. First, it was the impossibility of really evaluating the transformative potential of these experiences. Some of the more heart-wrenching accounts pointed to possibilities. However, others may also have had transformative potential, yet been more understated. The point here is that transformative experiences are hard to achieve and are probably rare. They are not easy to orchestrate.

Second, it is important to try. At the very least we can experiment with our pedagogy and construct activities that fall outside of the mainstream learning opportunities. Even if everyone does not have an overtly powerful experience, a breadth of educational opportunities can be provided. It might even be worthwhile to just hear first-hand accounts of the exceptional learning experiences of others.

Finally, when we undertake experiences such as those suggested here, we need to let go of ideas about control. To be transformative, probably means allowing for outcomes that cannot be outlined in advance. However, it is the sudden recognition of something extraordinary that transforms an individual, and we really do not know when this will happen. We can set up learning experiences that point in promising directions, and hope.

Næss's experience with a flea, and Leopold's experience with a wolf, could not have been planned or predicted. These just arose out of interesting contexts. Similarly, not everyone who works for an anti-poaching organisation has such a powerful experience as the South African blogger. Still, it seems that an educational imperative is to be as bold and creative as possible in imagining our pedagogies.

## LEARNING ACTIVITY 8.5: GOING BEYOND

When asked to do something imaginative, educators often point to heavy demands placed on them by prescribed curricula. They have a point. However, consider the alternative – that is doing nothing outside of the everyday expectations. Would anything change? It is our experience that exceptional educators will find ways to pry open little opportunities for more radical learning. That is, learning that digs a little closer to the roots of our larger and more systemic problems. It only takes a relative small number of them to make a big difference.

Sometimes, these educators frame their work as additive. Here the prescribed curriculum is the minimal learning and good teaching goes beyond doing just the minimum. Others describe their work in terms of resistance, or activism.

1.  In groups, discuss what could be done in your group that would be additive? Imaginative? Experiential? Activist? What could you do that would go beyond the minimum curriculum requirements? Remember that sometimes it can be more effective to take a lot of small steps, rather than large leaps.

2.  In pedagogical settings where outcomes are specified and tied closely to measurable evaluations, how could you present educational imperatives that don't neatly fit into prescribed learning outcomes? And are not easily evaluated? These are certainly challenging tasks, but they are important ones.

## SOME THOUGHTS ABOUT THEORY

*Reflections on modernity and the moral dimensions of proximity in ethics-led learning to change*

Many of the case stories shared in this chapter reflect an engagement with moral questions in a modern age of consumerism. These moral questions need to be carefully deliberated and carried through to exploratory action.

The case stories of the factory-farmed eggs and the zero-waste mean are only of value where they become "stories that do work" for us, enabling our critical reflections to be carried through into exploratory work that tells its own story and brings its own challenges. The activities that follow them may enable you to work with the stories to explore your own questions that arise around moral dilemmas and ethical expansion through proximity, reflexivity, and social innovation.

In all these cases, a close practical engagement brought a new dimension to the learning. A key dimension of this was how persons' moral impulses were clarified through a proximity experience, which in turn shaped their ethical purpose and became drivers of an expanding learning process.

Bauman notes how the modern world has become inherently hostile to proximity, a process that he sees can be contemplated as a crucible of moral action. He explores how, in the emergence of the modern state, the rules governing what is right are characterised by legislative and education processes that "compose and impose an all comprehensive, unitary ethics – that is, a cohesive code of moral rules which people could be taught and forced to obey."[18] These rules for governing social life came to be composed and taught through a proliferation of legislative and education interventions in the belief that all that was going wrong could be externally designed and brought into social life to put things right.

The impossibility of this task has us currently intermeshed amongst things that are going wrong in the world and in a reciprocal play amongst contested interpretations of ethics and reason. This perspective on the modern human condition notes that reason in modernity has had significant limits. And what is ethical cannot simply be legislated and implemented as an externality.

In the cases explored in this chapter, the learning engagement and ethical concerns have developed beyond earlier deliberations about moral problems and solutions. Here the analytical distance is closed through activities that are in proximity with real contexts, and where the intricacies of living issues come into play. These stories have tried to reflect a change in ethical outlook, and to show how an initial moral impulse developed into a continuing learning-led process of change.

For Bauman, proximity is a sustained condition of "being attentive and waiting" that must be enacted and repeated. In this way, individuals shape a latent state of being "for the Other". And through this sustained process of being in proximity with others, it may be possible for humans to shape an ethical social life in the company of others.

Moreover, seeing ourselves as more clearly in the company of others can become a shared search for liberating perspectives on a global scale. The projects described in this chapter attempt to do work in a different way – on behalf of all beings and within working global systems. It is also

---

[18]   Bauman, *Postmodern ethics*, 1993, 6.

notable that similar projects are likely to be unique in their own ways. Outcomes in each our cases can affirm the possibility that humans can go forward with positive practices in the face if seemingly intractable problems, and in some small ways can contribute to a wider common good.

## References

Bauman, Zygmunt. *Postmodern ethics*. Oxford: Blackwell Publishers, 1993.

Camus, Albert. *The Plague*. London: Penguin Classics, 2013.

De Leeuw, A. Dionys. "Contemplating the Interests of Fish." *Environmental Ethics* 18, no. 4 (1996): 373-390. https://doi.org/10.5840/enviroethics19961844

Jickling, Bob. "Ethics research in environmental education." *Southern African Journal of Environmental Education*, 22, (2005): 20-34. https://doi.org/10.1080/00958960309603496

Judt, Tony. "A hero for our times." *The Guardian*, November 2001, accessed July 17, 2019, https://www.theguardian.com/books/2001/nov/17/albertcamus

Kassi, Norma. "Science, ethics and wildlife management," in *Northern Protected Areas and Wilderness*, ed. Juri Peepre and Bob Jickling. Whitehorse, Yukon: Canadian Parks and Wilderness Society and Yukon College, 1994: 212-216.

Leopold, Aldo. A *Sand County Almanac: With Essays on Conservation from Round River*. New York: Sierra Club/Ballantine, 1966. First published in 1949/1953.

Næss, Arne. "Self realization: An ecological approach to being in the world," in *Thinking Like a Mountain: Towards a Council of All Beings*, eds. John Seed, Joanna Macy, Pat Fleming and Arne Næss. Gabriola Island, B.C.: New Society Publishers, 1988: 19-30.

Næss, Arne and Jickling, Bob. "Deep ecology and education: A conversation with Arne Næss." *Canadian Journal of Environmental Education* 5, (2000): 48-62.

Timmermans [Schudel], Ingrid and Lotz-Sisitka, Heila. "Learning through Environmental Policy Implementation: A case story of the Rhodes University Department of Education's environmental policy." *The Declaration*, 6, no. 2 (2003): 14-17.

# CHAPTER 9

## HOW CAN WE RE-IMAGINE THE FUTURE?

**Pedagogical intent:** So far we have been involved in various forms of critical analysis. But this is only half of our task. We criticise in order to envision real alternative possibilities. In this section, we take it as our task to actually re-imagine environmental and social possibilities.[1] This chapter illustrates that it is possible to see and do things differently and that it is possible to re-imagine the present, and to think about and propose alternatives that reflect a different or expanded ethics. The chapter explores inspiring and creative examples of re-imagining and re-framing and proposes a range of experiments that teachers can do with learners. We reflect on re-imagining activities at small-scale local level, but also how these can be expanded to include networks, movements, and systems.

**Activities:**    9.1: Re-imagining the future

                9.2: Reshaping the future through counter narratives

                9.3: Reshaping the future through self-validating invitation

                9.4: Reshaping the future through poetry

                9.5: Towards A dictionary of disruption: Re-jigging language to shape the world we want

                9.6: Reshaping the future through connective aesthetics

                9.7: Reshaping the future through sustained social practices

                9.8: Reshaping the future through expansive use of image, metaphor, and connective aesthetics

                9.9: Reshaping the future through social movements and networks

**Main theoretical resources:** At root here is Anthony Weston. In truth, it was him that first developed this workshop activity. For Weston, our greatest challenge is to create the space for environmental values to develop and evolve. We will importantly create new values by everyday practices by actively re-imagining the future. However, the new sources will bring new life into what we are doing, and we are excited about that! We share Appadurai's work on Hope and Social Movements; Bhaskar on absence and concrete utopia's; and De Sousa Santos and Vandenberghe on the formation of social movements involving absence and emergence, as well as reconstructive social theory. We surface Suzi Gablik's connective aesthetics, Beuys' social sculpture, and Egya's poetic activism.

---

[1]   This material is initially based on a workshop activity originally prepared by Anthony Weston. It also draws on stories from around the world published in the *New Internationalist Magazine.*

## Experiments in re-imagining the future

Marcuse[2] writes, "naming the things that are absent breaks the spell of things as they are." We practice critique – or deconstruction – ideally, in order to create a space to re-construct. In taking up our task of re-imagining social possibilities we claim the space opened by our critical work by trying to spell out – to make concrete – what we want instead. Similarly, Bhaskar argues for the importance of absence, as in this newly opened space, as a driver of change.[3]

In general, we know that we want to live at peace and as one with each other and with the more-than-human world, but how should we do this? How should we speak? What kinds of houses, policies, actions, events, enterprises, and relationships should we create? What sorts of stories should we tell our children? Importantly what is missing in the stories that we tell? How can these absences limit the scope of new stories that we could tell? We have barely begun to imagine the possibilities that exist. Yet, this too is an essential task for reconstructing cultural environmentalism.

We can begin by re-imagining images that we live by. United States philosopher Holmes Rolston III[4] spoke about visiting a favourite campground in the Rocky Mountains that is adjacent to sub-alpine meadows. The trail signs in these meadows, profuse with daisies, lupines, columbines, delphiniums, bluebells, paintbrushes, penstemons, shooting stars, and violets, for years read, "Please leave the flowers for others to enjoy". Later these signs were replaced by new ones saying, "Let the flowers live!".

A particularly powerful example of re-imagining the future comes from South Africa. Human rights abuses in that country were most sharply illustrated by the Sharpeville massacre of March 21, 1960 when police killed 69 people who were participating in a protest against Apartheid pass laws. This day had come to be commemorated as "Sharpeville Day". With the advent of democracy, the government retained the "day" but has re-named it "Human Rights Day". They reframed the issue in a way that could allow people to move forward *together*. Each March, South African people remember the past, through positive celebrations, and actions, chosen to celebrate a human rights culture.

So, what else can we do? One way to get started is to focus our work in three areas: Language, Social Practices, and Imagery. Later in this chapter we will also bring focus to social movements.

---

[2]   Herbert Marcuse, *One-Dimensional Man*.
[3]   Roy Bhaskar, *Dialectic: The pulse of freedom*.
[4]   Holmes Rolston III, "Ethics on the Home Planet," 107-139.

## Language

Some observers have noted that our value-language has become progressively narrowed and "one-dimensional"; ethical concerns have often been reduced to economics, or other singular, rationalist framings. For example, geological surveyors in Western Australia describe water as a resource for government-backed industry, in one instance for irrigating genetically modified cotton. By contrast, the Karajarri, the Aboriginal custodians of the same water, call it "Living Water", or *karnangkul*. They explain that Living Water is said to be inhabited by various "snakes" who are powerful beings that need to be respected. Not only are the words different, but so are the values and assumptions embedded in them. For the Karajarri people the language of water is integral to their cultural traditions and to their way of practicing law. To the geological surveyors the language of water conveys instrumental and industrial priorities.[5]

In another example David Lake[6] insightfully probes the language of climate change. Why do we talk, for example, of "global warming" as if it was something to make you cosy on a cold winter day? Perhaps a greater sense of urgency, or an "ethic of timeliness", would be better conveyed by "global heating", or even "hotting"?

In some cases re-naming could have powerful transformative and transgressive practical consequences.[7] In Aotearoa New Zealand an extraordinary "re-naming" process saved a river from unsustainable extractive industries, when the indigenous Te Awa Tupa Maori community not only reclaimed the Indigenous name of a river – The Whanganui River – but they re-named the river a "person" in legal terms.[8] In the Treaty of Waitangi, the community used what Cormac Cullinan[9] calls "wild law" to recognise the rights of a river to be a river, and recognising the river's rights to own itself. There is a kind of magic that happens when we use language to re-frame and re-narrate the world. In the case of Whanganui river, real tangible changes were made in the legal rights of a river. This case could have massive transformative effects on the global movement for the rights of nature.

Similarly, the 2010 Cochabamba declaration for the rights of Mother Earth[10] sets a precedent for other communities around the world to express their sovereignty. As a "renaming" process this declaration re-

---

5   Story adapted from Peter Yu, "Water fight."
6   David Lake, "Waging the War of the Words: Global Warming or Heating?" 52-37.
7   See Mick Strack, "Land and rivers can own themselves," 4-17.
8   Mihnea Tanasescu, *Environment, Political Representation, and the Challenge of Rights*, 107-128.
9   Referring to earth jurisprudence, and the recognising the rights of nature, see Cormac Cullinan, *Wild law: A manifesto for Earth Justice*, 52.
10  Evo Morales Ayma, Maude Barlow, Nnimmo Bassey, Shannon Biggs, Cormac Cullinan, Eduardo Galeano, Tom B.K. Goldtooth, Pat Mooney, Vandana Shiva, Pablo Solón, *The rights of nature.*

claims not only the name and language of the river, but the river's personal identity and capacity to own itself. This, too, could have global effects in transforming international environmental justice jurisprudence.

The challenge to us then is to *reclaim the language*, and to give words to values that are being denied in everyday conversations. It is also to reinvent a language to express the kinds of values we sense in the larger living world more precisely, but have not yet found the terms for? (See Learning Activity 9.5, "Towards a dictionary of disruption", for deeper exploration into reclaiming language to shape the world we want).

## Social Practices

Some say that the deepening of any kind of relationship with the more-than-human world is closed out by the way we build houses, schools, cities – in general, by how we live our lives. Imagine how we can redesign buildings that lessen the barrier between inside and outside – or make it more permeable. What would it mean to construct a building as if the outside mattered?

One interesting place to look for an answer to this question is a geographic location where the climate is challenging. Yellowknife, in Canada's Northwest Territories, can be characterised by its long cold winters – often an obstacle to embracing the permeability of living space. It is also notable for a cityscape that includes exposed rocky outcrops of Precambrian shield – that can be summertime gathering spots and hangouts. Interestingly, the designers of St. Patrick High School in Yellowknife chose to accept the challenge.

Figure 9.1   Foyer, St. Patrick High School Photo
(Photo by author, Bob Jickling)

It seems fortuitous that the building site included one of Yellowknife's rocky outcrops. Instead of blasting away the rock, the school entrance foyer was built around it. Facing a wall of windows to the outside, this rock now inside, serves as an assembly gathering spot – similar to those enjoyed in summer – and a lunchtime meeting place. Even on the coldest winter day there are visual and solid links to the outside. The school went further still. A core sample from "the rock" guided choices for paints, tiles, and carpets used throughout the school.

The policies and economic decisions that are made also affect the way we build relationships with others. Imagine, for example, if every country in the world put their military budgets towards providing education for all of the world's children, or making sure that all children have enough healthy, fresh food to eat. Imagine how social practices and social policies could be redesigned, so that instead of **self-validating reduction** they promote the opposite: something we might call "**self-validating *invitation***"?[11]

In Nicaragua, another example is the *Nueva Vida* (New Life) Women's Maquila. This cooperative was established by women and is operated by female workers who produce conventional and organic cotton clothes as an alternative to sweatshop employment. They were able to establish small markets in the United States, and in 2002 the group supplied 3000 T-shirts for a large European tour. Helped by a British trade-union initiative to wholesale non-sweatshop merchandise, these women have achieved dramatic changes in their lives.[12]

How about designing creative awards? Since 1980 the Right Livelihoods Awards are presented annually in the Swedish parliament to honour and support those offering practical and exemplary answers to the most urgent challenges facing us today.[13] The accompanying identifies some of the Environmental Laureates awarded through this programme.

---

[11]  These concepts, self-validating reduction and self-validating invitation, are introduced in Chapter 3 of this book.
[12]  Lenie Stockman, "Ethical threads," 6.
[13]  The Right Livelihood Award. "Laureates," http://www.rightlivelihoodaward.org /laureates/

# HANDOUT: ENVIRONMENTAL LAUREATES[14]

Greta Thunberg (2019, Sweden) …for inspiring and amplifying political demands for urgent climate action reflecting scientific facts.

Sheila Watt-Cloutier (2015, Canada) …for her lifelong work to protect the Inuit of the Arctic and defend their right to maintain their livelihoods and culture, which are acutely threatened by climate change.

Bill McKibben / 350.org (2014, USA) …for mobilising growing popular support in the USA and around the world for strong action to counter the threat of global climate change.

Hans Herren / Biovision Foundation (2013, Switzerland) …for his expertise and pioneering work in promoting a safe, secure and sustainable global food supply.

Nnimmo Bassey (2010, Nigeria) …for revealing the full ecological and human horrors of oil production and for his inspired work to strengthen the environmental movement in Nigeria and globally.

**Other interesting laureates:**

Syria Civil Defence (2016, Syria) …for their outstanding bravery, compassion and humanitarian engagement in rescuing civilians from the destruction of the Syrian civil war.

Mozn Hassan / NAZRA (2016, Egypt) …for asserting the equality and rights of women in circumstances where they are subject to ongoing violence, abuse and discrimination.

Svetlana Gannushkina (2016, Russia) …for her decades-long commitment to promoting human rights and justice for refugees and forced migrants, and tolerance among different ethnic groups.

Cumhuriyet (2016, Turkey) …for their fearless investigative journalism and commitment to freedom of expression in the face of oppression, censorship, imprisonment and death threats.

Tony de Brum & the People of the Marshall Islands (2015, Marshall Islands) …in recognition of their vision and courage to take legal action against the nuclear powers for failing to honour their disarmament obligations under the Nuclear Non-Proliferation Treaty and customary international law.

Kasha Jacqueline Nabagesera (2015, Uganda) …for her courage and persistence, despite violence and intimidation, in working for the right of LGBTI people to a life free from prejudice and persecution.

Gino Strada / EMERGENCY (2015, Italy) …for his great humanity and skill in providing outstanding medical and surgical services to the victims of conflict and injustice, while fearlessly addressing the causes of war.

Edward Snowden (2014, USA) …for his courage and skill in revealing the unprecedented extent of state surveillance violating basic democratic processes and constitutional rights.

---

[14]  Ibid.

## Imagery

Remember the images of nature described in earlier examples in this book. Here nature was presented as a problem or threat, as distant, as empty, as a commodity, and as an economic reduction of both the human and more-than-human worlds. All of this is concretely focused in metaphors such as the resource, playground, possession, or obstacle. Now ask: What would be the alternatives to this? What *new metaphors* do we need?

In an area where bears are still abundant, a shift has recently become visible. Warnings that once created images of bears as fearsome and dangerous creatures have been replaced by a message, "You are entering bear country" – you are the intruder, so be careful and respectful. How could we re-imagine signs in elephant habitat?

Figure 9.2    You are entering bear country (Photo of an interpretive sign at Liard Hot Springs in Northern British Columbia, by author, Bob Jickling)

What could be achieved if we thought of wild animals and plants – yes, even "weeds" – as neighbours, citizens, or original inhabitants?

How could priorities and assumptions be re-imagined if we stopped talking about the "developing" world, or "Third World" and started calling these "more sustainable" or "low footprint" countries or even the "Majority World" countries?

In each of the examples shared here, the imagery has, unlike the advertisements presented earlier, served to give presence to contradictions and controversy – or different stories. Each provides a starting point for re-imagining new possibilities.

# LEARNING ACTIVITY 9.1: RE-IMAGINING THE FUTURE

1. To begin, go outside in groups of two or three and find a place that is inviting – even with the snow, the heat, the dust, or, perhaps, because of them. Now imagine a way to invite others to become intimate with this place – a pedagogy of intimacy. What can you come up with? What activities could you introduce? How could you help to make the experience a self-validating invitation?

2. Now, in your same groups, choose one, or maybe two, of the three dimensions of this re-imagining topic and work hard to seriously imagine possibilities. Be prepared to share your ideas. Consider:

   **Language** – What language is used in the environment you live or work in? Can you *reclaim language*, or *reinvent a language custom* to adequately express the kinds of values you sense in the larger living world, perhaps, but have not yet found – or are being denied – the terms for. So, what kinds of other concepts/words do you need? Propose some! Make a case for your approach.

   **Social Practices** – What are the prevailing, or dominant, social practices that shape your environment? Try to imagine how social practices and social policies can be redesigned, so that instead of self-validating reduction they promote the opposite: "self-validating *invitation*". What kinds of houses, schools, and cities ought we to build? What kinds of work should there be? What kinds of policies will be needed? What kinds of rhythms should the places have? What kinds of teaching practices can we develop? Again, be concrete and specific in your proposals.

   **Images** – What are the prevailing images in the environment you are in? What new metaphors do we need? How would they be made concrete? When a politician or local school principal waves the new flag, what should he or she say? Show?

## Counter narratives

The concept of counter-narrative is particularly interesting when reclaiming and re-imagining language. By this concept, we mean the process of taking one story line that has traditionally reflected a set of social norms and rewriting it to frame a positive alternative. For example, consider the role of the "*#Pussyhatproject*" during the Women's March at the same time as Donald Trump's inauguration. In addition to manifesting a social movement, it was also a good example of a counter narrative. That is, this project took the derogatory and disempowering use of the word "pussy" by Trump and then reclaimed the word for another purpose. That is, the counter-narrative was in making the word respectable again when used by women to delegitimise misogynist norms.

Johanna Ferreira,[15] in her essay "Reclaiming the word 'pussy' in a post-Trump era," speaks about how a word has been disempowering and negative. She reminds us how this word is often used to describe a man as being "weak" or "wimpish," or to objectify and subject women, predominantly in popular film, music, and pornography. On the 21st January 2017, "pussy" was transformed into a word synonymous with power and strength. Slogans on placards, t-shirts, hats and banners now read "PUSSY POWER".

Figure 9.3    Pussy hats worn by protesters during the 2017 Women's March (Photo courtesy of Sallie A. Reissman)

## LEARNING ACTIVITY 9.2: RESHAPING THE FUTURE THROUGH COUNTER NARRATIVES

We are reminded of an older gentleman who, when accused of being a "tree hugger", would respond: "Yes, the long-armed variety." This was his way of disarming a scoffing critic and embracing the term "tree hugger". Similarly, the term "green" or "greenie" has been used in disparaging ways, but now is often presented as a harbinger of new ways of doing things. (Though we acknowledge that "green" is also used in cynical and exploitive ways, too.)

1.  In small groups, brainstorm examples from your region where words have been reclaimed. That is, examples where a word that can have

---

15    See Johanna Ferreira, "Reclaiming the word 'pussy' in a post-Trump era."

negative connotations is being used in a positive way as a means of highlighting a problem. What examples can you find?

2. Can your group come up with a storyline, or word, that they would like to like to "take back?" Can you think of an individual project that would be worth taking on? Or maybe there could be a school-wide project to generate a counter-narrative? What would it be?

## Self-validating Invitation

Self-validating invitation is another kind of counter narrative. However, in this case it can have the capacity to become self-reinforcing. The more that people consider the invitation, the greater it becomes, in a perpetuating cycle. This phenomenon can also be seen in opposition to self-validating reduction, as described in Chapter 3.

Self-validating reduction is akin to the concept of self-fulfilling prophecy familiar to many teachers and other educators. Here, for example, if teachers believe that a student is a poor performer, that student may as a result take on the characteristics of a poor performer, thus reinforcing the original belief. This can become a self-reinforcing process as the teacher's beliefs and interactions with the student become increasingly negative.

Here we pose the question: Can this cycle of reduction be countered? And what would self-validating invitation look like? How might it be enacted?

A simple example might come through a discovery in your natural environment. Walking along the muddy shore of a river, or maybe snowshoeing in fresh snow, might reveal the tracks of our animal neighbours – often the ones we rarely see, like wild cats, or wolves, or nocturnal critters. Once aware of these spoor, we can become more attentive. We are thus invited to see the landscape as inhabited by more-than-human neighbours – ones that we seldom see directly. The more we notice this cohabitation, the more likely we are to pick up on additional cues – tracks, sounds, and maybe even sightings. The more that we tune into the clues around us, the more we see. As this experience repeats itself over time, we can be invited to see a place as a diversely inhabited more-than-human world.

Taking these ideas further, we describe some possibilities for self-validating invitation found in two complex issues. In the first case, we return to the killing of Cecil the lion. This case was first presented in Chapter 3 as an example of self-validating reduction. However, over time Cecil's death provoked other reactions. Consider whether you believe these responses indicate the beginnings of something like self-validating invitation.

The second case describes a project begun by the Namibian Ombetja Yehinga Organisation. They were searching for new ways to teach about HIV and AIDS.

## Killing Cecil the Lion

In Chapter 3 we reproduced a news story about the killing of Cecil the lion by an American hunter. A learning activity in this chapter asked whether the news reporting contained messages and metaphors that might have reflected a kind of self-validating reduction of animals in general, and Cecil in particular. By this we mean the kind of reduction that can, for example, lead people to see animals as simply commodities for human amusement. However, we can also ask whether the response to Cecil's killing has generated a counter-narrative that is the beginning of a self-validating invitation?

We can consider this question by looking at the international response to the killing of Cecil. One contemporary gauge of public response to contentious issues can be found through postings on Wikipedia. Of course, the medium is a living source of information that is continually expanding and is subject to self-correcting forces. Still, in just over eight months following Cecil's death, the following excerpts were noted:

> The killing drew international media attention and sparked outrage among animal conservationists, politicians and celebrities, as well as a strong negative response against Palmer. Five months after the killing of Cecil, the U.S. Fish and Wildlife Service added two subspecies of lion, in India and Western and Central Africa, to the endangered species list, which includes the species of Cecil, making it more difficult for US citizens to kill these lions. According to Wayne Pacelle of the Humane Society, Cecil had "changed the atmospherics on the issue of trophy hunting around the world," adding "I think it gave less wiggle room to regulators."[16]

> Several celebrities publicly condemned Cecil's killing. Artists from around the world dedicated art to Cecil, including former Disney animator Aaron Blaise.

> The killing of Cecil sparked a discussion among conservation organisations about the ethics and business of big-game hunting and a proposal for bills banning imports of lion trophies to the USA and European Union. These discussions have convinced three of the largest airlines in the USA – American, Delta, and United – to take the voluntary step of banning the transport of hunting trophies. In response, under the premise that profits from trophy hunts help animal conservation efforts, Pohamba Shifeta, the Namibian Minister of Environment and Tourism said, "This will be the end of conservation in Namibia." Activists have also

---

16   Wikipedia, "Killing of Cecil the lion."

called on African countries to ban bow-hunting, lion baiting, and hunting from hunting blinds. Global media and social media reaction has resulted in close to 1.2 million people signing online petition, "Justice for Cecil", which calls on Zimbabwe's government to stop issuing hunting permits for endangered animals:

> Safari Club International responded by suspending both Palmer's and Bronkhorst's memberships, stating that "those who intentionally take wildlife illegally should be prosecuted and punished to the maximum extent allowed by law." Late-night talk-show host Jimmy Kimmel helped raise US$150,000 in donations in less than 24 hours to Oxford's Wildlife Conservation Research Unit, which had been "responsible for tracking Cecil's activity and location."[17]

It is early to say whether the outrage over Cecil's death will really change "the atmospherics on the issue of trophy hunting around the world", or meaningfully disrupt any new views that might be becoming entrenched in particular societies. The prospects of new views emerging are undoubtedly more complex that the snapshot presented here. Still, these excerpts do suggest that concerted resistance to unpopular trends can disrupt their uncritical acceptance, and that new outlooks can arise and begin to counter ongoing self-validating reduction.

## The Caring Namibian Man

The Namibian Ombetja Yehinga Organisation felt that conventional methods were not particularly effective for teaching about HIV and AIDS, and that a more creative response was needed. Their experiment, which can be interpreted as an example of self-validating invitation was called "The Caring Namibian Man". They have now developed into a Trust that works to engage young people in the task of challenging social practices and assumptions with creative artistic experiments.

As authors, we were inspired by one of their projects that culminated in a touring art show called "The Caring Namibian Man". This project began by recognising that African men, especially during the current AIDS epidemic, are often characterised as aggressive, abusive, and misogynistic. You can imagine how such stereotypes could easily lead to self-validating reduction of a huge segment of the population. While this may be true for many men, it is wrongly generalised and it does little to break the cycle of violence. Seeking to disrupt this cycle and promote alternatives, the Ombetja Yehinga Organisation distributed cameras to a group of youth and challenged them to find and photograph examples of "caring" Namibian men. In other words, they were invited to find and celebrate men who disrupted this stereotype. The result was a moving collection of

---

17    Ibid.

photographs that were exhibited around Southern Africa. An example from this show is provided below.

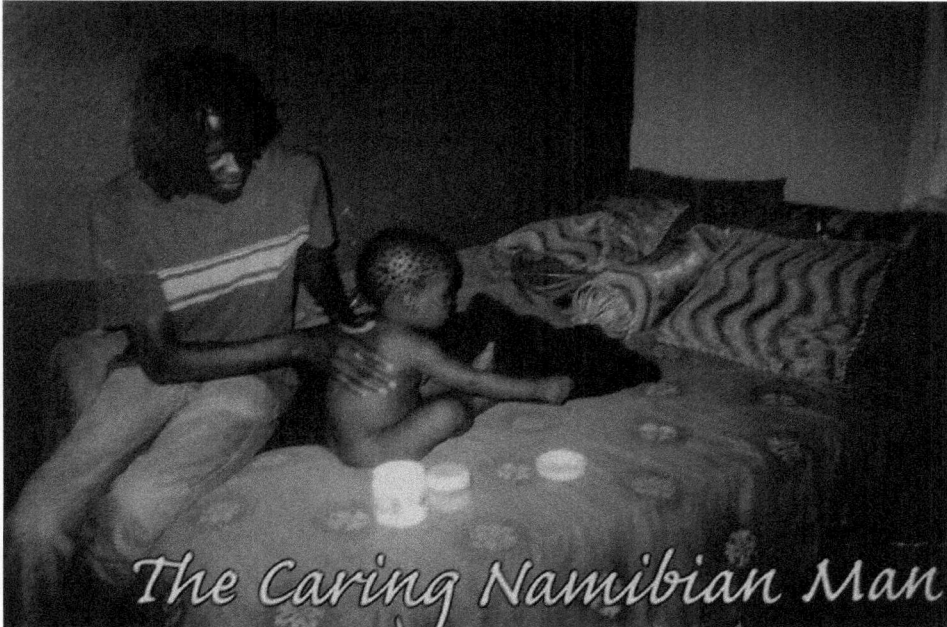

Figure 9.4    The caring Namibian (Image courtesy of the Ombetja Yehinga Organisation Trust)[18]

## LEARNING ACTIVITY 9.3: RESHAPING THE FUTURE THROUGH SELF-VALIDATING INVITATION

1. Consider the description of the Caring Namibian Man project and the picture above. Now discuss the stereotypes or assumptions about men that are challenged by this project. How does this project challenge assumptions about what a "real man" is? Think about how these assumptions about real men can contribute to self-validating reduction. Now, how does this project illustrate a process that could be called self-validating invitation?

2. Think about Africa's "big five" game animals – the lion, the elephant, the rhino, the buffalo, and the leopard. These five charismatic and "larger than life" mammals are central characters in popular children's films and books and the focus of gripping documentary series, and are considered a "must-see" on any safari trip in Africa. Of course, they are

---

[18]   More information about this organisation and "The Caring Namibian Man" project can be found at: www.ombetja.org

also legendary to big game hunters. At times, it can seem that Africa's wildlife has been reduced to these five species.

In an effort to show the rest of the world that Africa has an incredible biodiversity and to draw attention to smaller and less noticed animals of the savannah, children in schools and tourists are now introduced to the "little five". These are the ant lion, the elephant shrew, the rhino beetle, the buffalo weaver, and the leopard tortoise.

Figure 9.5   Caring for dung beetles in South Africa
            (Photo by author, Bob Jickling)

Signs that warn the park visitors not to drive over elephant dung often surprise tourists to the Addo Elephant Park in South Africa. As it turns out, this dung is a nutritious food source and is often densely populated by dung beetles. These beetles also gather the dung to make their brood nests. Can you see the self-validating invitation in this sign?

What other examples can be found that could be interpreted as self-validating invitation?

## EXTENSION ACTIVITY

3. Now look around you for a more challenging project. What self-validating reductions can you find? How can you take this destructive cycle and invent a project that challenges the reductions with a self-

validating invitation? Best of all, find a self-validating invitational project that you and your group can actually implement.

## Poetry

Poetry has a long activist history. It has developed and thrived as an oral tradition around the world. It is enacted in spoken form and in song, as well as in written form.

In Africa, Sule Emmanuel Egya[19] explores the use of poetry and the reclaiming of language in the Niger Delta by Nigerian poets and others. The delta has been under siege from various multinational extractive oil corporations where dirty practices poison rivers, clear forests, and generally displace communities from their homelands. It is through using what Egya[20] calls "metaphor and nuanced metaphoric and metonymic aesthetics" that they enliven and awaken thinking and knowing. Put simply this means the ability to change and transform and repurpose language and concepts for activism and transformation. A common example of metonymic aesthetics is the way in which activists are re-emphasising the word "responsibility" as "response-ability". This is the ability to respond: to reframe deeply what we mean when we ask people to be responsible for the environment's wellbeing. Responsibility moves from being an individual obligation with imperatives, towards an acknowledgement of structures and contexts in which the individual responds. Egya saw this as a way to enliven and awaken thinking and knowing beyond the status quo through what he refers to as the "poetics of activism":

> It is also that poets from the Niger Delta region, in particular, see themselves as literary militants, engaging in what one might see as a poetics of activism to combat institutional powers that perpetrate slow violence leading to both spiritual and physical displacements.[21]

The poets Egya examines include people, such Gabriel Okara, Christian Otobotekere, Tanure Ojaide, Ogaga Ifowodo, Nnimmo Bassey, and Ebi Yeibo.

The idea of metonymic aesthetics is useful in thinking about re-imagining and reclaiming language, as metonymic speaks to the almost sculptural process of substituting and renaming a concept or idea (or in the case of the Niger delta poets – an injustice) with a physical object or place. It is a way of reclaiming not just the words, but the meaning and the cultural being into the land and the environment. As Egya explains for the Niger Delta poets, the act of writing poetry itself is, is a process of resisting, reclaiming, and re-existing – it is seen by the more radical poets as:

---

19   Sule Emmanuel Egya, "Nature and Environmentalism of the Poor," 1-12a.
20   Ibid., 2.
21   Ibid., 8.

a form of activism … the same sense that the militants, as many of them argue, struggle for the liberation of the region. In other words, poetry becomes a powerful instrument for conveying resistance; it historicizes the condition of the peoples and their lands, and by doing so raises a counter-narrative in confrontation against powerful institutions, governmental and otherwise, on behalf of the poor people and their environment.[22]

## LEARNING ACTIVITY 9.4: RESHAPING THE FUTURE THROUGH POETRY

1.  In small groups, discuss what poetic forms are most represented in your area? In spoken word or song? Are there activist poets that speak to the concerns and fears of your group? How do such activist poets affect the choices you make and the things you do?

2.  Go online and read some work of the poets Sule Emmanuel Egya. Better still, listen to them where audio recordings and/or videos are available? Can you feel the difference between reading poetry and hearing it read or sung?

3.  See if you can attend poetry readings in your community. If so, why not give poetry a try? Maybe attend a few readings and see if you can detect an activist thread running through the work.

## LEARNING ACTIVITY 9.5: TOWARDS A DICTIONARY OF DISRUPTION: RE-JIGGING LANGUAGE TO SHAPE THE WORLD WE WANT

Taryn Pereira-Kaplan, a poet/activist in South Africa, has written a T-learning Tiny Book for activists called "Towards a dictionary of disruption".[23] Here she asks if we could repurpose words to shape the world we want:

Consider, for example, a few conceptual re-jigging's of powerful words and phrases:

- Why do we always label "Indigenous people" as separate and different – we should surely rather have a labeling name for the colonisers, the newcomers? (like Pākehā, the Maori language term for New Zealanders of European descent).
- Let us stop talking about "invisible knowledge" and rather turn our language lens on "those who lack the ability to see."

---

22  Ibid., 11.
23  Taryn Pereira-Kaplan, "Towards a Dictionary of Disruption."

- "There is really no such thing as the 'voiceless.' There are only the deliberately silenced, or the preferably unheard."[24]

Taryn asks us to "…collect and develop these repurposed words and concepts, and use them to shape the world we want.

- Think of words that you would like to repurpose to use to shape a situation you feel needs changing.
- Perhaps share this activity with your peers, classmates or your students, and create a dictionary of disruption for your context.

## Going further: Looking at sustained re-shaping of the future

The preceding section introduces the idea that we can, and indeed must, re-imagine the future. This includes how it will sound and look, and what we might do. There are many profound experiments documented and presented together as a kind of broad overview.

In this section we will look a little more deeply at a few examples that animate this spirit of re-imagination. Indeed, these examples seem to have moved beyond isolated experiments and morphed into social movements. As you move into this phase of our book you might consider the question: Are you able to lift out what inequalities are being responded to, and begin to see how they are interrelated?

### Connective aesthetics

Can you imagine making change without trying something new? In pondering this question, we are aware that much historical work in environmental ethics is weighted in favour of linguistic and logical kinds of analysis and expression. However, we should probably ask, are these tools broad enough to capture sufficiently new perspectives that can actually provoke change? The philosopher and poet Jan Zwicky[25] invites us into a space opened by such questions.

In important ways, Zwicky brings us back to questions and challenges raised in Chapter 5. There we began to reach beyond strictly logical and linguistic approaches to understanding our presence in the world. Here, environmental ethics broke from tradition and – pointed to something new – by recognising that we are physical beings whose embedded values are linked to the things we do, how we have lived, our connections to the land, and the relationships we build.

Importantly for this chapter, Zwicky suggests that when philosophy attempts to give voice to ecologies of experience – to understand how things will affect us as physical beings with emotions and to know things that exist outside of human language and logic – we are doing something

---

24  Arundhati Roy, "Peace & The New Corporate Liberation Theology," 4.
25  Jan Zwicky, "What is lyric philosophy?" 3-18.

different. For her, when we try such things, it is no longer useful to distinguish between art and philosophy.[26] This, of course, opens a huge palette of possibilities, some of which build on activities described in Chapter 5. In this chapter we use the opening to introduce art making that gives importance to "connective aesthetics" and art practices that attempt to build social and environmental connections.

In 1992 the artist and art critic Suzi Gablik wrote a seminal commentary, "Connective Aesthetics" in the journal *American Art*.[27] In this article she embraced what was seen as an emerging new paradigm in art – one that affirms relatedness and connection rather than an ideal of freedom and individuality that was, at the time, dominating art practice. In ways that seem resonant with Zwicky, she sees an important turning from abstract thinking in art towards more concrete expressions of being in the world and this is the basis of connective aesthetics.

Gablik sees this turning as a yearning for a sense of community and intimacy, but also an urge to revise guiding cultural myths. For her, art practice, as in other aspects of society, had become driven by subtexts of power, profit, and patriarchy. This, she suggests is not a very creative response to our planet's needs – then, or for that matter, now. She goes on to suggest that in an ecological world view, "the self is no longer isolated and self-contained, but relational and independent".[28]

For Gablik, the politics of connective aesthetics was different from the more conventional art practice of her time. For her differences are at least twofold. First, connective aesthetics makes art more socially and environmentally responsive. Art embraces and inhabits environmental and social practice. Second, it challenges us to renew our collective being in the world. It calls us to see through the eyes of others – human and more-than-human. It calls us to stretch our boundaries beyond the ego-centric self to create a wider view of the world, and of ourselves. It calls us to be a different and larger notion of self.

In this larger notion of self that Gablik talks about, we find resonance with Arne Næss's notion of the large "S" Self discussed in Chapter 5. In both instances they were proposing something radically different. Næss was suggesting an inverting of the order of environmental ethics, and Gablik was doing the same in the art world. Together, we see Zwicky's tendency to see a fusion of art and philosophy.

In the end Gablik predicts that that in the decades to follow there would be more art that would be essentially social – connective and collaborative. This has proven true. Indeed, the top prize at the prestigious Venice Biennale in 2019[29] was awarded to Lithuanian artists working in this

26   Ibid.
27   Suzi Gablik, "Connective Aesthetics," 2-7.
28   Ibid., 4.
29   Farah Nayeri, "Venice Biennale's Top Prize Goes to Lithuania."

tradition. More expansive views about connective aesthetics are developed later in this chapter.

For now, though, we will focus on an activity to help educators to get started.

## LEARNING ACTIVITY 9.6: RESHAPING THE FUTURE THROUGH CONNECTIVE AESTHETICS

Teacher education students were given the broad task of re-imaging the future during the cold of a Canadian winter. This often necessitated a first step in re-imagining winter itself. That is, they needed to think about way to make being outside more inviting – for people to be more connected to a winter environment.

The task was framed in an artistic way and students were encouraged to think of snow as their palette, and to work collaboratively. Some projects took the form of snow sculptures, often with a cultural challenge. For example, snow televisions were sometimes placed beside busy pathways to provoke discussions about social practices.

In another instance, a class spotted a large pile of snow placed to the side of a freshly cleared parking lot. They chose to sculpt a circular outdoor classroom on its summit. Subsequently, their class used this site to hold class discussions in their ongoing practices of reconnecting with a winter environment.

In your area what would make a good palette for a collaborative outdoor project designed to connect students with their place and with each other? With this in mind, frame an activity that you could use in your own teaching setting that would embrace the idea of connective aesthetics.

### Sustained Social Practices

**Reducing Boundaries.** Efforts to reduce the boundaries between inside and outside in an effective and sustained way will be important in any social practices that seek to deepen relationships with the more-than-human world. A poignant example of making that boundary more permeable comes from Stephen Ritz, a high school teacher in the South Bronx district in New York, USA. Ritz and his students re-imagined their classroom as a space to grow fresh food: initially for their class, and eventually the entire school. Over time, this project was transformed into a wider movement that included healthy eating among students and their families.

They farmed plants and vegetables indoors by creating "edible walls". In turn this strategy gave rise to the "Green Bronx Machine," a project to help other schools in the USA "start their own agricultural programmes to teach children healthy eating, environmental awareness and life skills."[30]

---

[30] Matthew Jenkin, "My edible classroom gives deprived New York kids a reason to attend school."

Ritz continues his educational work in New York and travels the world promoting the value of growing fresh produce, both in schools and the wider community. Ritz reflects on the project:

> I wound up working at a very troubled high school.... It had a very low graduation rate and the bulk of my kids were special educational needs, English language learners, in foster care or homeless. It was dysfunctional to say the least. Someone sent me a box of daffodil bulbs one day and I hid them behind a radiator – I didn't know what they were and figured they may cause problems in class. A while later, there was an incident in the room; we looked behind the radiator and there were all these flowers. The steam from the radiator forced the bulbs to grow.... That was when I realised that collectively and collaboratively we could grow something greater. We started taking over abandoned lots and doing landscape gardening, really just to beautify our neighbourhood. We took forlorn, unproductive spaces and turned them into aspirational places where we could bring communities together. We then moved on to growing food indoors in vertical planters around the school.... By building an "edible wall" to grow fresh vegetables in our science classroom I gave the kids a reason to come to school. Everyone likes to be outside getting their hands dirty, but I needed kids to be in school. I realised that if I could control a lot of the variables – such as watering, pests and weeding – by growing plants indoors, gardening would be even more fun.[31]

Figure 9.6    Ritz and his student tending to their edible wall[32]

---

[31]    Ibid.,
[32]    Image sourced from Matthew Jenkin, "My edible classroom gives deprived New York kids a reason to attend school." Original photo source, Progressive Photos.

Here a teacher and his students collaboratively transformed their classroom with the unexpected result of transforming their school and neighbourhood. This re-imagining and "making-with" process also created a sense of belonging and purpose for the students who before felt alienated and disconnected from the classroom. As Ritz reflects:

> The kids really believe that they are responsible for them and attendance has increased from 43% to 93%. Students come to school to take care of their plants – they want to see them succeed. Along the way, the kids succeed too. That's great, because if I have their bodies in school, I have their brain. I am committed to the belief that children should not need to leave their neighbourhood to learn and earn in a better one. Why not take the urban setting they live in and turn it into an oasis?[33]

***

A similar transformative, yet nomadic learning space, emerged from the back alleys and dirt roads of Argentina. Activist-artist Raul Lemesoff, created what he calls "Arma De Instruccion Masiva," or "Weapon of Mass Instruction". Although sounds rather ominous, it is really a peaceful mobile library. Lemesoff's creation can be classified as a hybrid between an artwork/vehicle/public library. Here he transforms the violent concept of "weapon" into a creative space for learning.

What is particularly significant about his mobile library is that it resembles a military tank. He took a 1979 Ford Falcon and repurposed it as a "tank" that carries approximately 900 books rather than soldiers and artillery. The library offers free reading materials to the public, in a variety of urban and rural spaces. Raul considers the library as a way "to contribute to peace through literature.[34]"

As Charley Cameron reports: "Where the vehicle once brought military oppression, it now brings literature of all genres in a collection constantly replenished through private donations."

---

[33] Matthew Jenkin, "My edible classroom gives deprived New York kids a reason to attend school."

[34] For Raul Lemesoff's "Weapon of Mass Instruction," see Charley Cameron, "'Weapon of Mass Instruction' is a Mobile Library."

Figure 9.7    Raul Lemesoff behind the wheel of his "Weapon of mass instruction"[35]

Lemesoff's art, that is transforming a weapon into a library, is a powerful image that opens up the tremendous opportunity there is to re-imagine the role the military could serve in society and create space for freely accessible education. Inherent is this approach to art making is what is sometimes called a "connective aesthetic". That means maintaining deeply connected relationships with society and places during the artistic process. This idea will be taken up further in the next section.

**Intentional communities.** It seems suitably celebratory to return to Anthony Weston at this time. Anthony penned the first versions of "Re-imagining the Future". He has been doing just that for decades. Fittingly, one of his more recent projects has been working towards establishing a farm-based community in central North Carolina.

Affectionately known as Hart's Mill Ecovillage,[36] this intentional community is becoming a reality on 112 acres of fields, forests, and streams. The members' hope that Hart's Mill will contribute to reversing ecological degradation across environmental, social, and economic dimensions. As a rural village, they aspire to develop a close community with space to roam for people of all ages and walks of life.

**Transition towns.** For a larger scale project, there is the international Transition Network and Transition Towns, Networks and Hubs. As their website declares, this is "a movement of communities coming together to re-imagine and rebuild our world".[37]

---

[35]  Image sourced from Charley Cameron, "'Weapon of Mass Instruction' is a Mobile Library." Original photo source "Image (cc) by Carlos Adampol on Flickr."
[36]  Common Ground Ecovillage, www.hartsmill.org
[37]  See Transition Network, www.transitionnetwork.org

## LEARNING ACTIVITY 9.7: RESHAPING THE FUTURE THROUGH SUSTAINED SOCIAL PRACTICES

1. It is interesting, particularly in Stephen Ritz's example, that transformative educational experiences can often have quite simple and seemingly innocuous beginnings. This gives credence to the importance of experimenting with new ideas and new pedagogies. What ideas or teaching strategies have you wanted to try? What would it take for you to get them started? Can you develop an implementation plan?

2. Art is often on the cutting edge of activism, even when that does not seem to be the explicit goal. It may be that artists are often gifted with an ability to see the world through many lenses – most of which diverge from the status quo? What would it take to nurture this gift? Develop a learner-oriented art project that encourages multiple perspectives on your neighbourhood, community, country – or go global? Or maybe develop an art project that builds on relationships with place and society.

   In conceiving this activity, feel free to step beyond conventional thinking about art. As Raul Lemesoff shows us, the most important aesthetics can often be found in the art making processes and in experiences in places.

   More examples of place-based art projects can be found in Chapter 5.

## EXTENSION ACTIVITY

3. In this section we have taken a more in-depth look at examples of social practices that began as small projects and developed into large and sustained points for community engagement. Seek examples of such sustained activist projects in your own community. Are there any intentional communities? Where is the nearest transition town, or network? What can they bring to your learning?

   Develop ways to bring them into you teaching and learning. Do they provide suitable opportunities for field trips? If so, consider organizing an outing.

Of course the idea of activism and education can pose dilemmas for educators. For ideas about navigating these dilemmas see Chapters 3 and 11. Of course there are no easy or correct answers. Rather, these chapters may provide some touchstones for consideration.

## Expansive use of image, metaphor, and connective aesthetics

The role that imagery and metaphor play in connecting people to ideas, history, place, and culture cannot be underestimated. It is at the heart of connective aesthetics.

Within the etymology of the word "metaphor", we see its origins in the Greek "to transfer or transport" which are synonymous with the contemporary Greek word for "truck" – therefore metaphor is a means of transport.[38] We could extend the translation further to also refer to the power of image to "transgress" the boundaries of our thinking and being. The use of image/metaphor in poetry, visual arts, theatre, and storytelling in general transports us into unknown lands and imagined futures.

The artist Joseph Beuys explored this expansive idea of image, metaphor, and aesthetic in profound ways in his expanded theory of art and practice. He called this "Social Sculpture" and "Warmth Work". In helping us to see where he is going with this, Beuys also spoke of the "the warmth character of thought" that can be seen as the warmth that softens fat or wax.[39] These ideas manifest in his famous project called Honey Pump in the Work Place, where he worked with the images and physical aesthetics around him – in this case honey and fat – as a connective force to shape the ways in which we come together to learn and plan in our world.[40]

Beuy's Honey Pump was installed around the staircase of the Fridericianum Museum, in Kassel, Germany. It consisted of a series of tubes running into rooms adjacent to the staircase, through which two tons of liquid honey was pumped by a margarine-lubricated motor. For 100 days, Beuys and many others who came and went, used the setting created by this installation to engage in a "permanent conference" as part of his Free International University project. They gathered there each week to explore different agendas and proposals through a detailed deliberative democratic process. All the while the honey moved through three different levels, from the basement, in the central conversation space, and up into the ceiling.

Beuys used the constantly moving honey to embody: "the circulation of capital ... an organic, material circulation, or an organic circulation of money like a human being's blood circulation ..." Thus, the honey pump enabled a "connective practice" that encouraged a unique form of social exchange and possibilities for learning.

The honey pump, like a heart, enabled an emergent atmosphere and imaginative inspirations in the 100-day permanent conference. As Beuys described:

---

38  Daniel W. Conway and Peter S. Groff, Nietzsche: critical assessments.
39  See for example Joseph Beuys, "Eintritt in ein Lebewesen"; and Shelley Sacks, "Social Sculpture and New Organs of Perception: New practices and new pedagogy for a humane and ecologically visible future," 80-98.
40  See for example Joseph Beuys, "What is Art?" in What is Art? 46.

one can say the heart with its circulation represents a sensing, feeling, movement principle, and the head is the form principle, then the will element was still missing. So, with this machine one could say: all three important creative factors, the three most important structures were represented: thinking, feeling and movement and the power of the will. One didn't need to be familiar with such concepts or be able to identify them. One could just experience them; many people experienced them.[41]

It is also possible to say that this installation together with the "permanent conference" revealed possibilities for "warming-up" a space and making it more pliable or plastic – receptive to different agendas.

In German "plastik" refers to both sculpture and the pliable plastic transformability of matter and human imagination. Therefore, Beuys's theory of sculpture was a theory of social plasticity – of pliability and transformability. Here he seems to amplify Susi Gablik's concept of "connective aesthetics" and reminds us of the original power of art process and art-making as a force that connects us to concepts or ideas that might just be out of reach.[42]

The psychoanalyst Carl Jung carefully revealed the power of image and symbol in human consciousness in meaning making and being.[43] Beuys runs with this idea in his famous remarks: "Every human being is an artist, a freedom being, called to participate in transforming and reshaping the conditions, thinking and structures that shape and inform our lives …"[44]

More recently, artist and educational researcher, Dylan McGarry has built directly upon the work of both Suzi Gablik and Joseph Beuys.[45] In his project he created a mobile social-learning platform on a train that traversed South Africa. Here the aim was to create a 44 day "permanent conference" in the spirit of Beuys as an alternative to traditional "conference space."

McGarry's project was also conceived and created through a collaborative social movement of "cultural practitioners" ranging from visual artists, poets, filmmakers, theatre-makers, guerrilla-gardeners, musicians, facilitators, to educational researchers, among others. The train was a practical response to windowless rooms of UNCCC climate negotiations which seem to exclude the people most deeply affected by climate change, who were the least responsible for causing it. The train was a re-imagining of the conference itself, and rather than being in a fixed place, it was a traveling filled with creative people who had created

---

41  Ibid.
42  Suzi Gablik, "Connective Aesthetics," 2-7.
43  Carl Gustav Jung and Marie-Louise Franz, *Man and his symbols.*
44  Joseph Beuys, quoted in Sandy Nairne, *The State of the Art.*
45  Dylan McGarry, "The Listening Train," 8-21.

receptive artworks and processes to deeply listen and empathise with the real-world challenges facing South Africans.

Methodologically, the experiment builds upon Gablik's sense of "connective aesthetics" to establish new ways of thinking, working, and reflecting together. The train also allowed for a re-imagining of social spaces, and a reframing of the narratives of climate change. It also created a space where groups could co-define the issues that were of the most concern for them, develop code of practice that could guide how they wanted to work together and also enabled practitioners on the train, and ability to work from a morally intuitive space rather than one driven by rules and imperative.

# LEARNING ACTIVITY 9.8: RESHAPING THE FUTURE THROUGH EXPANSIVE USE OF IMAGE METAPHOR AND CONNECTIVE AESTHETICS

1.  Consider one of the examples described above (McGarry's *Listening Train*, or Lemesoff's *Weapons of Mass Instruction*). In what way did the connective aesthetic facilitate solidarity and parity (inclusiveness)? Why do you think images, sculptures, stories, poems unite people, or bridge divides?

2.  Consider how using an art-based process and a connective aesthetic process could contribute to your work? Begin with what Joseph Beuys' asked himself often: "what needs warming up?" i.e., what feels cold, static, or disconnected in your situation/work/context, and what image, story or artifact could be used to bridge the disconnection. It could be a small creative act or action (connective aesthetic). In the next section you will see how the simple small act of knitting a pink hat, started a revolution.

## Social Movements and Networks

Sometimes experiments in language use, social practices, and creative imagery morph and grow into social movements and networks with widespread followings and significant social impact.

An example of an emerging social movement is represented in the striking image of millions of pink "cat-eared" hats transforming the streets of Washington into sites of protest during a women's march on the day of United States President Trump's inauguration. What can only be described as a rose-tinted-warm-woollen-ocean of pro-women activism was a profound sight to behold and created an equally moving aesthetic for the Women's March. Here was an aesthetic that helped connect women and unite them in their cause.

The hats were a satirical response to Trump's audio recorded comment "I moved on her like a bitch … Grab them by the pussy. You can do anything" that he later tried to dismiss as "locker-room banter". What soon became known as *The Pussyhat Project* was initiated by Krista Suh and Jayna Zweiman in Los Angeles. Unable to attend the march themselves, they decided to create handmade pink woollen hats to be worn at the march for visual impact, and they encouraged others to do the same.

The response to their call-to-knit was overwhelming. Crafters from all over the USA began knitting, crocheting, or sewing "pussy hats" using patterns provided by the Suh and Zweiman. This simple hat-sharing gesture became a profound social learning space and cultural exchange.

What is remarkable about this case is that Suh and Zweiman put out their call less than two months prior to the march. They simply wanted to participate, they had an idea to share, and they invited others to participate. They could not know that their project would become a social movement. What they did not expect was that on the day of Trump's inauguration over 600 different demonstrations were recorded across the USA, with around 3-4 million participating in the USA, and around 5 million participating globally.[46] Pussyhats were seen in all these locations, even including a demonstration in Antarctica. The Pussyhats made it to the cover of TIME magazine with the headline: "The resistance rises: When a march became a movement."[47]

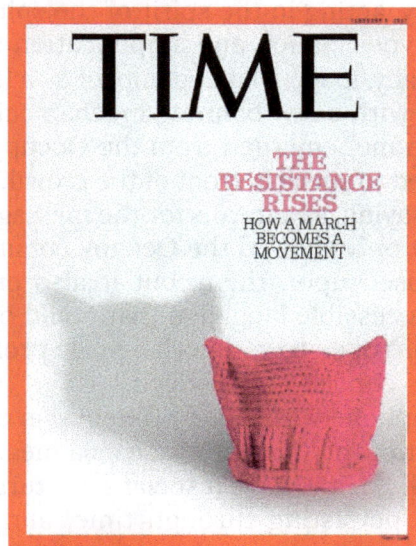

Figure 9.8    "The resistance rises: When a march became a movement."[48]

---

[46]    Wikipedia. "2017 Women's March."

[47]    Karl Vick, "Perhaps the Largest Protest in U.S. History Was Brought to You by Trump."

[48]    *Time Magazine*, Online Cover, February 6, 2017, accessed February 13, 2021, https://time.com/vault/year/2017/ . Photo credit, Danielle Amy Staif, for *Time*.

Still, at the beginning, the Pussyhat resistance was simply an idea of a connective aesthetic that was shared along with patterns useful in duplicating the hats. In hindsight, the idea was captivating and the timing was perfect.

Sometimes, relatively simple ideas have the capacity to generate strong and pervasive social movements with increasingly complex responses to social and environmental issues. The Occupy Wall Street movement provides a profound contemporary example of how this works.

The Occupy Wall Street movement emerged as a response to the social and economic inequality, greed, corruption, and the undue influence of corporations on government – particularly from the financial services sector. The Occupy Wall Street slogan, "We are the 99%", refers to income and wealth inequality in the USA between the wealthiest 1% and the rest of the population. To achieve their goals, protesters acted on consensus-based decisions made in general assemblies which emphasised redress through direct action over the petitioning to authorities.[49] The connective aesthetic of "occupation" transferred to a global social movement where protesters went on to occupy banks, corporate headquarters, board meetings, foreclosed homes, and college and university campuses.

Erika Biddle[50] explores how Beuys' Free International University and his honey pump imagery can be linked to the General Assembly formed in the early days of The Occupy Wall Street movement. According to Biddle, this General Assembly, acting in the spirit of the spirit of the university, "served as a form of vocalization and amplification for the movement's participatory democracy, made a lasting impact as a learning site."

When confronted with a law banning mechanically amplified sound, "the people's microphone" emerged from the Occupy Wall Street space. When someone wanted to speak in front of the crowd, the crowd repeated back what they were saying, as a means for the message to spread through large groups. This not only allowed the Occupy community to transgress the boundaries imposed upon them, but it also created a culture of speaking in simple, accessible language that could be repeated, making the cultural space more open to various language proficiencies, as well as keeping conversations concise and distilled.

The people's microphone was, indeed, a Beuysian connective aesthetic tool that emerged out of the Occupy space as a means to overcome real challenges in the now. So, a very real social sculptural process emerged. An artistic aesthetic persisting through time, and amplified through the Occupy Wall Street Protests, thus contributed to the rise of a powerful movement.

---

49   See Berrett, "Intellectual Roots of Wall St. Protest Lie in Academe."
50   Erika Biddle, "Re-Animating Joseph Beuys' 'Social Sculpture'," 29.

## LEARNING ACTIVITY 9.9: RESHAPING THE FUTURE THROUGH SOCIAL MOVEMENTS AND NETWORKS

Not all great ideas become social movements, but they do all begin with inspired activity. Do you have an activity that you have always wanted to share? Would this be a good time to try it out? Change often begins with a few dedicated people taking small actions. Many examples of social movements originate around social justice issues, as in the Pussyhat example.

1. Do some research in your community. Are there environmental issues that have coalesced into community-scaled social movements, or networks? Or is there a chapter of a larger movement or network in your community? Is there a movement that you would feel comfortable joining? What criteria would you use to make such a decision?

2. Can you think of an issue where social and environmental issues are linked? Such complex situations could be a worthy starting point to begin acting to build a network of connective activists.

3. What are the most pressing and most dynamic social movements and networks you are aware of in your neighbourhood, region, or country? What common identities and social categories are working together among these different movements that produce advantages (or disadvantages) for human and more-than-human communities? Are you able to lift out what inequalities are being responded to, and how they may be interrelated?

## SOME THOUGHTS ABOUT THEORY

Like education, ethics is an idea that is used in a variety of ways. It is often framed in terms of questions like: What is a good life? What is a good way to live? What should I do? How should I live? As processes of inquiry these questions can be consistent with the conception of education. However, this often comes with a fear that "if you let ethics off the rational leash, it will turn into ideology."[51] For anxious people, there is always the fear that someone else's values, or code for conduct, will be imposed on their children. With this concern about misplaced ideology, pedagogical responses to ethics need to be considered carefully in light of conceptions of education. You could say that approaches to ethics and education are more likely to gain traction if they are copacetic.

Anthony Weston's paper, "Before environmental ethics,"[52] suggests that environmental ethics is still in its "originary" stages; it is evolving and we still have little idea where it will take us. He argues that it is far too early to foreclose discussion about theoretical perspectives, converge on particular assumptions, and outline rules for conduct. For Weston, environmental ethics is in need of a great deal of exploration. We should expect, at best, a long period of experimentation and uncertainty. Like Naess and Bauman, Weston is not looking for an ethical framework, or moral codes to implement. Rather, he appears to be interested in a more imaginative, and process-oriented approach to ethics.

In an interesting elaboration, Weston argues that our challenge is not to systematise environmental values, but rather to create the "space" for environmental values to evolve. By space he is speaking about the social, psychological, and phenomenological preconditions that are needed to enable this evolutionary process. He is also speaking about the conceptual, experiential, and physical freedom to move and think. Here Weston is concerned that individuals and groups *can* begin to create, or co-evolve, new values through everyday practices. Our job might thus be seen as "enabling environmental practice".[53] For Weston, too, there is an essential role for life's experiences in his vision for ethics.[54]

At this point we see important points of coherence between ethics and the ideas about education. Weston argues convincingly that environmental ethics is still evolving and that there is no need to adopt a particular set of assumptions, ethical framework, or ideology. Indeed, that would be counterproductive to the continued development of the field. Here we

---

[51] John Ralston Saul, *On equilibrium*, 86.
[52] Anthony Weston, "Before environmental ethics," 321-338.
[53] Ibid.
[54] There are of course many others that have similarly argued in favour of contextualising environmental ethics in place-based experiences. Also consider the pioneering work of Jim Cheney, "Postmodern environmental ethics," 117-134; Val Plumwood, "Nature, self, and gender," 3-21; and Karen Warren, "The power and the promise of ecological feminism," 125-146.

have an approach that vigorously opposes letting ethics slip into some ideology. Rather, ethics is framed as a process of reflection, imagination, and experimentation where individuals and/or groups create new ways of being in their part of the world. This strikes us as a practical way to approach ethics that can be educationally justifiable. If education is about doing the seemingly impossible, then imagination and experimentation will be infinitely more appropriate than flirting with dogma and doctrine.

## Practical adequacy

We started this book by talking about stories and their capacity to do ethical work. This is different from the exercise of developing ethical theories that seek to identify basic ethical truths and then to develop frameworks for universally applying these truths in a wide range of contexts. We are still waiting for a perfectly usable theory, which may well be beyond our reach given the complexity and the diversity of the world. Yet we know that there is more to ethics than this; humans ultimately are intimately related to, are situated within, and are part of, the more-than-human world. This complexity provides the impetus for us to seek out ways of thinking about what is practically adequate when it comes to moral life.

Knowledge of the world, or our beliefs and values about the world, and the world to which it refers may not be the same thing. As history shows us, our knowledge, beliefs, and values relating to the world differ in time and space. For example, Stone Age astronomers might have had very good predictive understanding of the stars and developed beliefs and values around this knowledge, but they may have not known what stars are from a physics perspective.[55] Today astronomers and physicists have more knowledge of what stars are, and may even have more sophisticated predictive knowledge of our relations with the stars and have developed more secular ways of relating to and valuing the stars. However, in comparison to future generations' needs or knowledge capabilities, this knowledge and this belief and values system may as yet be wholly inadequate, for all we know.

Yet it is possible to say that the knowledge of and beliefs surrounding the stars in each period has had (or may have) practical adequacy, in relation to the time and space in which it is generated and used. This requires reflexivity. As Andrew Sayer says,

> We are always in some position or other in relation to our objects; the important thing is to consider whether that influence is benign or malign.[56]

---

[55] This example of our knowledge of the stars is adapted from Andrew Sayer, *Realism and social science*.

[56] Ibid., 53.

Andrew Sayer helps us think this through by saying that while there is no escaping engaging in normative theory or making moral choices and judgements, we can try to be careful how we do this. He suggests that to achieve practical adequacy in our normative or moral engagements we can try to avoid 1) **empty moralising** in which we seek to provide judgement from a "moral high ground" that fails to engage adequately in the world with the issues and concerns; and 2) **avoiding a naive position** that assumes that whatever is agreed to be good will come into being. He suggests that we need to do additional work, and question whether normative goals are *feasible*, not only in terms of whether we can get from A to B, but also in terms of whether B is feasible in the first place. Here, he also suggests that it can be helpful to think about "utopias", but not impossible utopias that are unfeasible or impossible to attain. We need to think about possible or concrete utopias practically.

In doing this work, we do of course need to think normatively about good and bad forms of social order, and good and bad forms of social-ecological relations. Without this, we could not engage with ethical discussions or practices. Describing and explaining the forms of social organisation or social-ecological relations that exist, and those that could feasibly exist at some other time (recognising that this might not be a fully accurate description), is a helpful starting point. We believe that this is illustrated through the diverse activities in this book.

We propose reconsidering theoretical work in ethics as stories – stories that tell us about the forms of social organisation or social-ecological relations that exist, and social-ecological relations that could feasibly exist at some other time. This does not demand absolute agreement about the truth of various theoretical premises, or the universal applicability of their ethical frameworks; rather, working with theoretical work in ethics in this storied way helps us to describe and explain the forms of social-ecological relations that exist, or that could feasibly exist, so that we can search out "practical adequacies" that offer some hope for the future and the mutual flourishing of all. We ask, despite flaws and irreconcilable conundrums that the history of ethics theorising has produced, do these stories still carry sufficient weight to allow us to do ethical work? We must make judgments in everyday issues and these ethical stories can help.

## Social movements and networks

The power of imagery and connective aesthetics is astonishing, especially at the scale at which a simple image such as a pink *"pussy-hat"* can become the symbol of the largest march in USA history. A pink hat that transformed a march into a movement and generated new social spaces in which people are able to struggle and work through democratic violence and the tyranny of oppressive political regimes.

What social movements are teaching us about re-imagining action is not just the power of poetic and aesthetic action, but also the power of how we re-frame our narratives, and in so doing reframe the way we engage with others – particularly how we reframe our relationship with oppressors. The struggle at Standing Rock reservation (also known as the Dakota Access Pipeline Protests)[57] and the #NODAPL movement opens a kind of relational and intersectional action that is emerging in critical Environmental Justice movements more broadly.

First, we see the brave efforts of people at Standing Rock reclaiming language by calling themselves "water protectors" as opposed to activists or protesters. They were also clear to steer away from language that associated them with larger bigger conversations around global climate change, as they were cautious of their efforts becoming a small "sound bite" or token of the global climate justice community. So, their careful re-framing of language and how they identified themselves carved out a unique niche and position in the greater extract-ivism struggle.

Second, in an essay on the "changing of hearts" at Standing Rock, Charles Eisenstein explores the way in which the Elders encouraged the youth to invite oppressors – in this case the police, and hired security, and eventually military – off the "warpath" through appealing to their shared humanity and common-ness:

> When you have a chance of overcoming an opponent by force, then fighting is a reasonable option. Absent that condition, victory has to come some other way: through the exercise of a kind of power that makes guns, money, and other kinds of coercive force irrelevant. Dare we call this power love?[58]

Eisenstein then shows how the people at Standing Rock used song, prayer, ceremony, and nonviolent resistance to absent violence and offer alternative forms of communication. The Elders guided younger water protectors to disrupt the narratives that legitimise the forceful suppression of the Water Protectors – synonymous with narratives about violent extremists, criminal elements, protecting the public, and so forth. By using

---

[57] The Dakota Access Pipeline protests, also referred to as hashtag #NoDAPL, is a collection of grassroots movements that began in early 2016 in response to the approved construction of Energy Transfer Partners' Dakota Access Pipeline in the northern United States. The pipeline was projected to run from the Bakken oil fields in western North Dakota to southern Illinois, crossing underneath the Missouri and Mississippi Rivers, as well as under part of Lake Oahe near the Standing Rock Indian Reservation. Many in the Standing Rock tribe (and other citizens) consider the pipeline and its intended crossing of the Missouri River to constitute a threat to the region's clean water and to ancient burial grounds. In April 2016 Standing Rock Sioux elder LaDonna Brave Bull Allard established a camp as a center for cultural preservation and spiritual resistance to the pipeline; over the summer the camp grew to thousands of people, and water protectors came from across the country and stayed through the winter.

[58] Charles Eisenstein, "Standing Rock: A Change of Heart."

these non-violence activities at Standing Rock, the Elders explained, the government could not forcefully evict peaceful Water Protectors. They reasoned, if the protesters were violent to the police, it would justify the police's use of violence against them.

According to Eisenstein violence is pre-empted by a process of dehumanisation of the other that we see in in war, racism, and genocide. For the Elders at standing rock, it was no different to any reduction of the sacred to the profane. In so doing the elders were very careful to ensure what language was used when speaking to law-enforcement personnel. They reportedly asked water protectors to proceed "prayerfully". That is to be aware of the sacred, and what could be called sacred speech, and reverence for the other, even in the height of struggle.

It was this prayerful impulse that opened intersectional relationships between war veterans and the Water Protectors. By December 2016, hundreds of United States military veterans joined the Water Protectors. In a video statement to the BBC, Michael Wood Jr., a former USA Marine veteran and Baltimore police officer, said:

> If the cops there want to be state-sanctioned agents to brutally beat non-violent veterans, that have served their country honourably – if they're going to beat us – then that should be the signal to the rest of the world of what our country's doing.[59]

It is these invitations off the warpath that opened relationality and intersectional solidarity. They presented opportunities to re-frame the discourse through absenting the enabling narratives of violence that are all too familiar in public protests around the world.

Charles Eisenstein sees the Standing Rock example as an opportunity to employ what Gandhi called "soul force". Where we meet violence with nonviolence. This counter-invitation has the potential to invite new dialogical spaces akin to the "dialectical" pulses of freedom that Roy Bhaskar discusses. It is manifest, in a sense, through the absenting of violence through song, careful use of poetic language, and through mutual reverence.[60]

There are murmurings that give us a sense that Gandhi's "soul force" is not only re-emerging in Indigenous people's movements but also in the intersectionality[61] between other movements responding to structural and institutional "Othering." The Black Lives Matter movement, Queer

---

[59] BBC News, "Standing Rock: US Veterans Join North Dakota Protests."
[60] Roy Bhaskar, *Possibility of naturalism*, 58.
[61] Intersectionality is a concept aimed to describe the ways that many identities and social categories work together to produce advantages and disadvantages across bodies and space, and that inequalities do not act independently of one another. See for example, Patricia Hill Collins, *Black Feminist Thought*; Grace Kyungwon Hong, *Ruptures of American Capital*.

and gender movements, and environmental justice movements are also awakening to shared "soul force".

David Pellow[62] carefully reveals this growing relationship and how Othering and responses to Othering unify these different movements, and their agency. He explores how the Black Lives Matter movement has revealed the mechanisms of structural racism that drives abnormal police shootings of young African American men, and how these same structural forces of Othering are at play against the more-than-human world. The Black Lives Matter movement re-frames these institutionalised and structural narratives by creating what Pellow refers to as racial indispensability, which is a re-framing narrative that purposefully contradicts:

> ... the ideology of White supremacy and human dominionism and articulating the perspective that excluded, marginalized, and othered populations, beings, and things – both human and more-than-human – must be viewed not as expendable but rather as *indispensable* to our collective futures.[63]

In this way, racial indispensability informs what Pellow calls socio-ecological indispensability (when referring to broader communities within and across the human/more-than-human divide and their relationships to one another).

The framing of indispensability is deliberately aimed to challenge the institutional, policy, and practical blind spots that support and perpetuate anti-Black racism. These blind spots share the same flawed assumption that the future of African Americans is somehow de-linked from the future of White communities. The idea of indispensability is distinct and transformative of the liberal assimilationist perspective, of "we are all one" which seeks to – often involuntarily and violently – incorporate "others" into one's own vision of a society.[64] Rather, Pellow argues, the concept and expression of indispensability honours fundamental environmental justice and ecological principles by recognising all communities – more-than-human and human – as interconnected, interdependent, but also sovereign and requiring the solidarity of Others.

What we are re-imagining in contemporary social movements is not only how we can share a sense of common ecological citizenship, but also an ability to express our unique and sovereign identities, within the commons. De Sousa Santos[65] echoes Pellow when he offers us an especially useful metaphor when he recognises the "inexhaustible and ungraspable diversity of social experiences in the world" and describes them as a

---

[62] David Pellow, "Toward A Critical Environmental Justice Studies," 221-236.
[63] Ibid., 231.
[64] Pellow citing Andrea Smith, *Conquest: Sexual Violence and American Indian Genocide*.
[65] Boaventura De Sousa Santos, "A Non-Occidentalist West? 112.

living, organic, and infinitely diverse "ecology of knowledge". He locates our ecology of knowing or knowledge(s) within liberation movements – particularly those responding against colonialism and the Othering of normative capitalist patriarchy. Here he includes feminism; ecology; Indigenous movements, the Afro-descendent civil rights movement; the peasant movement; liberation theology; the urban movement; the Lesbian, Gay, Bisexual, Transgender, Intersex, Asexual, Queer, Two Spirited (LGBTIAQ2S) movement; and describes how these movements have birthed new "conceptions of life and human dignity, new symbolic universes, new cosmogonies … and even ontologies."[66] While De Sousa Santos reveals this infinite diversity of human experience and knowledge, he shows how this emergent narrative has revealed a contradictory realisation – that this infinitude of experience exists in the "finitude of the planet earth, the unity between humanity and the nature inhabiting the world … and the limits of life sustainability on earth."

## References

Ayma, Evo Morales, Maude Barlow, Nnimmo Bassey, Shannon Biggs, Cormac Cullinan, Eduardo Galeano, Tom B.K. Goldtooth, Pat Mooney, Vandana Shiva, and Pablo Solón. *The rights of nature: The case for a universal declaration of the rights of mother earth.* Ottawa: Council of Canadians, 2011.

BBC News. "Standing Rock: US Veterans Join North Dakota Protests," accessed December 2, 2015, http://www.bbc.com/news/world-us -canada-38188624

Berrett, Dan. "Intellectual Roots of Wall St. Protest Lie in Academe: Movement's principles arise from scholarship on anarchy," *The Chronicle of Higher Education*, October 16 2011, accessed February 23, 2012, https://www.chronicle.com/article/intellectual-roots-of-wall-st -protest-lie-in-academe/

Beuys, Joseph. "Eintritt in ein Lebewesen," Lecture given during Documenta 6 in Kassel, Germany, 1977. Audio cassette. Wangen: FIU.

Beuys, Joseph. "What is Art?" in *What is Art? Conversations with Joseph Beuys*, ed. Volker Harlan. East Sussex: Clairview Books, 2004.

Bhaskar, Roy. *Dialectic: The pulse of freedom.* London: Verso, 1993.

Bhaskar, Roy. *Possibility of naturalism.* 3rd Edition. London: Routledge, 1998.

Biddle, Erica. "Re-Animating Joseph Beuys' 'Social Sculpture': Artistic Interventions and the Occupy Movement." *Communication and Critical/Cultural Studies* 11, no. 1 (2014): 25-33. https://doi.org/10.1080 /14791420.2013.830810

Cameron, Charley. "'Weapon of Mass Instruction' is a Mobile Library That Disseminates Free Books in Argentina." Inhabit (2012), accessed

---

[66] Ibid.

February 13, 2021, http://inhabitat.com/weapon-of-mass-instruction-is
-a-mobile-library-that-disseminates-free-books-in-argentina/

*Canadian Journal of Environmental Education*, 11 (2006).

Cheney, Jim. "Postmodern environmental ethics: Ethics as bioregional narrative." *Environmental Ethics* 11, no. 2 (1989): 117-134. https://doi.org /10.5840/enviroethics198911231

Collins, Patricia Hill. *Black Feminist Thought*. New York: Routledge, 2008.

Conway, Daniel W and Groff, Peter S. ed. Nietzsche: critical assessments (Vol. 1). New York: SUNY Press, 1998.

Cullinan, Cormac. *Wild law: A manifesto for Earth Justice*. Cape Town, South Africa: Siber Ink, 2002.

Culver, Carolyne. *New Internationalist*, January 5 (2002): 29.

De Sousa Santos, Boaventura. "A Non-Occidentalist West? Learned Ignorance and Ecology of Knowledge." *Theory, Culture & Society* 26, no. 7-8 (2009): 103-125. https://doi.org/10.1177/0263276409348079

Egya, Sule Emmanuel. "Nature and Environmentalism of the Poor: Eco-poetry From the Niger Delta Region of Nigeria." *Journal of African Cultural Studies*, 11, no. 1, (2016): 1-12a. https://doi.org/10.1080/13696815 .2015.1083848

Eisenstein, Charles. "Standing Rock: A Change of Heart." *Charles Eisenstein*, December 2, 2016, http://charleseisenstein.net/standing-rock -a-change-of-heart/

Ferreira, Johanna. "Reclaiming the word 'pussy' in a post-Trump era." *thecnnekt*, last modified March 1, 2017, accessed July 16, 2019, http:// www.thecnnekt.com/stories/2017/2/23/reclaiming-the-word-pussy-in -a-post-trump-era

Gablik, Suzi. "Connective Aesthetics." *American Art* 6, no. 2 (1992): 2-7. https://doi.org/10.1086/424147

Galeano, Eduardo. *Upside Down*. New York: Metropolitan Books, 2002.

Global Alliance for the Rights of Nature. Universal Declaration of Rights of Mother Earth. 2010. *The right of Nature*, accessed September 1, 2018, http://therightsofnature.org/universal-declaration/

Hong, Grace Kyungwon. *Ruptures of American Capital: Women of Color Feminism and the Culture of Immigrant Labor*. Minneapolis, MN: University of Minnesota Press, 2006. https://doi.org/10.5749/j.ctttvbtk

Jenkin, Matthew. "My edible classroom gives deprived New York kids a reason to attend school." *The Guardian*, August 20, 2014, accessed February 13, 2021, https://www.theguardian.com/teacher-network /teacher-blog/2014/aug/20/classroom-attend-school-fresh-food -healthy-eating-students

Jung, Carl Gustav and von Franz, Marie-Louise. *Man and his symbols*. London: Aldus Books, 1964.

TheJournal.ie. "Just eight days of military spending could make education free around the world," July 12, 2015, http://jrnl.ie/2202883

Lake, David. "Waging the War of the Words: Global Warming or Heating?" *Canadian Journal of Environmental Education* 6 (2001): 52-37.

McGarry, Dylan. "The Listening Train: A Collaborative, Connective Aesthetics Approach to Transgressive Social Learning." *Southern African Journal of Environmental Education*, 31, no.1 (2016): 8-21, www.ajol.info/index.php/sajee/article/view/137658/127221

Marcuse, Herbert. *One-Dimensional Man*. Boston: Beacon Press, 1964.

Nairne, Sandy. *The State of the Art: Ideas and Images in the 1980s*. London: Thames and Hudson, 1987.

Farah Nayeri, "Venice Biennale's Top Prize Goes to Lithuania," *New York Times*, March 11, 2019, accessed July 16, 2018, https://www.nytimes.com/2019/05/11/arts/venice-biennale-winner.html?searchResultPosition=1

Wikipedia. "Killing of Cecil the lion," accessed March 21, 2016, https://en.wikipedia.org/w/index.php?title=Killing_of_Cecil_the_lion&oldid=906184345

Pellow, David N. "Toward A Critical Environmental Justice Studies: Black Lives Matter as an Environmental Justice Challenge." *Du Bois Review: Social Science Research on Race* 13, no. 2 (2016): 221-236. https://doi.org/10.1017/S1742058X1600014X

Pereira, Taryn. "Towards a Dictionary of Disruption". *T-Learning Tiny Book*. 2018. http://transgressivelearning.org/resources/

Plumwood, Val. "Nature, self, and gender: Feminism, environmental philosophy, and the critique of rationalism." *Hypatia* 6, no. 1 (1991): 3-21. https://doi.org/10.1111/j.1527-2001.1991.tb00206.x

Pretty, Jules and Hine, Rachel. *Reducing Food Poverty with Sustainable Agriculture: A summary of New Evidence*. Final report of the SAFE-World Project, Centre for Environment and Society. Essex, England: University of Essex, 2001.

Rolston III, Holmes, "Ethics on the Home Planet," in *An Invitation to Environmental Philosophy*, ed. Anthony Weston. New York: Oxford University Press, 1999, 107-139.

Roy, Arundhati. "Peace & The New Corporate Liberation Theology," The 2004 City of Sydney Peace Prize Lecture. Sydney, Australia: The Centre for Peace and Conflict Studies, University of Sydney, Occasional Paper No. 04/2, November 3, 2004.

Sacks, Shelley. "Social Sculpture and New Organs of Perception: New practices and new pedagogy for a humane and ecologically visible future," in *Beuysian Legacies in Ireland and Beyond: Art Culture and Politics*, ed. Christa-Maria Lerm Hayes and Victoria Walters. Berlin: Lit Verlag, 2011, 80-98.

Saul, John Ralston. *On equilibrium*. Toronto: Penguin, 2001.

Sayer, Andrew. *Realism and social science*. London: Sage Publications, 2000. https://doi.org/10.4135/9781446218730

Smith, Andrea. *Conquest: Sexual Violence and American Indian Genocide.* Boston, MA: South End Press, 2005. https://doi.org/10.1515/9780822374817

Strack, Mick. "Land and rivers can own themselves." *International Journal of Law in the Built Environment* 9, no. 1 (2017): 4-17. https://doi.org/10.1108/IJLBE-10-2016-0016

Stockman, Lenie. "Ethical threads: Nicaraguan women's co-op offers alternative to sweatshops." *New Internationalist*, December 2003, 6.

The Right Livelihood Award. "Laureates," accessed July 16, 2019, http://www.rightlivelihoodaward.org/laureates/

Tanasescu, Mihnea. *Environment, Political Representation, and the Challenge of Rights.* London: Palgrave Macmillan, 2016, 107-128. https://doi.org/10.1057/9781137538956_6

Warren, Karen. "The power and the promise of ecological feminism." *Environmental Ethics* 12, no.2 (1990): 125-46. https://doi.org/10.5840/enviroethics199012221

Weston, Anthony. "Before environmental ethics." *Environmental Ethics* 14, no. 4 (1992): 321-338. https://doi.org/10.5840/enviroethics19921444

Wikipedia. "2017 Women's March," accessed July 16, 2019, https://en.wikipedia.org/wiki/2017_Women's_March

Wikipedia. "Killing of Cecil the lion," accessed March 21, 2016, https://en.wikipedia.org/w/index.php?title=KillingofCecilthelion&oldid=906184345

Vick, Karl. "Perhaps the Largest Protest in U.S. History Was Brought to You by Trump." *Time*, January 25, 2017, accessed September 12, 2018, http://time.com/4649891/protest-donald-trump/

Yu, Peter. "Water fight: Aboriginal law and agricultural development plans clash," *New Internationalist*, July 1999: 314.

Zwicky, J. "What is lyric philosophy?" in *Alkibiades' love: Essays in philosophy*, ed. Jan Zwicky. Montreal & Kingston: McGill-Queens University Press, 2015, 3-18.

# CHAPTER 10

## How elusive is ethical action?

This chapter proposes a series of questions to help us to think about our options and choices for ethical action. These questions are aimed to help guide or direct our actions in ethically considered ways that are right for individuals or groups in particular situations at particular times and places. At the same time, these choices need to be possible in terms of planetary boundaries and other cultural, social, and economic opportunities and limitations, considerate of all alternatives, and morally justifiable. Such justifications need to be based on the basic needs and rights of ourselves, others, and the more-than human world.

**Activities:**              10.1: "Fairing" free trade?

                            10.2: Exercising choice as producers and consumers

                            10.3: Ethical actions for rhinos

**Main theoretical resources:** The main resource for this chapter is Bhaskar's dialectical critical realism which searches for a way of describing being and becoming in a world beset by injustices and inconsistencies. The chapter is based on the knowledge that there is no way we cannot act and so highlights the importance of exploring ethical ways of acting – or acting through non-action – in the world.

## How can we act in an ethically sound way?

The previous chapters of this book have emphasised the role of ethical debates in helping to guide our actions, especially in the quandaries of Chapter 4. It can often be difficult to choose an appropriate ethics-informed, or value-formed action. There is always more than one way forward. Another layer of complication arises when we consider the socio-cultural setting in which we make our choices, as highlighted in Chapter 6. These settings influence both how we feel about things but also our agency in effecting change.[1]

This chapter builds on work from earlier chapters in imagining what actions are possible and morally defensible. First, however: What does it mean "to act?"

The philosopher, Roy Bhaskar argues that: "You cannot not act. You must act. If you abstain from acting, that too is an action is it not? That is an action, that is choice."[2] We always have agency in a system whether our

---

[1]    Val Plumwood, "Paths beyond human-centeredness."
[2]    Roy Bhaskar, From science to emancipation, 307.

actions oppose or support the status quo. If the former, acting in opposition to the status quo, our actions can push directly against oppressive norms or they can work indirectly in provocative ways.

A powerful example of provocative indirect action is in the story of the Nobel Peace laureate Leymah Gbowee who helped organise a "sex strike" amongst a group of women. The action of withholding sex was instrumental in ending a long-running civil war in Liberia.[3] Actions can also include direct confrontations such as employing the legal system to challenge certain oppressive actions or inequalities. For example, direct action is present in cases where industry contravenes environmental law or threatens the health of workers and surrounding communities. Another example of a direct action would be cycling or walking to work instead of taking your car to reduce climate change emissions. An indirect action might be canvassing the municipality to establish cycle paths to make the roads safer for cyclists and pedestrians – thus "paving the way" for reduced carbon emissions.

This chapter examines agency, particularly the moment of deciding what to do, and then doing it. Many questions at this point are: How can we act in a morally defensible way? How do we settle on an "ethical action"? How can we be considerate of ourselves, of others and of the more-than-human world? What if we are faced with conflicting emotions, contradictory understandings, or "inconvenient" truths?

If we want to find a way forward, we cannot simply accept all truths, or all actions, as having equal validity – that is, we cannot take an ethically relativist stance. Environmental educator Leigh Price tells us that when faced with contradictions we cannot just agree to disagree. By doing so we fail to decide on a way forward.[4] When conflicts arise, accepting many truths leaves us incapable of agreeing on a collective way forward. Thus, in this chapter we seek "grounds for preferring one belief … to another"[5] and one action to another.

We do this by working with different case studies, looking at different environmental conundrums, and examining the different actions people and groups have taken in response to those quandaries. We think about them and ask: Can we call these ethical actions, that are morally justifiable and the best options in a particular context and a particular time?

In order to deal with the complexity of ethical engagements in the world, we propose a set of thinking tools for helping to guide these engagements. The learning activities are designed to introduce these tools.

---

3   Democracy Now. "Liberian Nobel Peace Prize laureate Leymah Gbowee."
4   See Leigh Price, Transdisciplinary explanatory critique of environmental education.
5   Roy Bhaskar, Possibility of naturalism, 58.

## LEARNING ACTIVITY 10.1: "FAIRING" FREE TRADE?

This activity employs a series of questions that serve as thinking tools to analyze a particular issue. In this case, the questions are about free and fair trade, although the questions could easily apply to other issues.

The handout "What is Fair Trade?" gives some insight into the economic theory of free trade and a proposed alternative – fair trade. Small groups can be given the following questions to discuss after reading this handout:

1. What **inconsistencies and contradictions** arise in practice? For example, you might like to think about: How "free" is free trade? Can you spot any examples of **oppressive relations** in your analysis of free trade? Does free trade illustrate practices that oppose the rights and freedoms of humans and/or the more-than-human world?

2. What emancipatory relations does fair trade offer that could enable different ways of "being" in the world? Think about the practice of fair trade. Is it either **challenging or supporting the dominant economic system**, that is, the economic *status quo*?

3. In the practice of free trade how do individuals' **relationships with themselves and others – including the more-than-human world –** influence the choices they make?

4. What are some of the **implementation challenges** that might arise for particular people in particular places and times in establishing fair trade policy and practice?

Now, as a whole group, bring your thoughts together. Discuss your findings and consider whether free trade can be considered an "ethical action" or not. What about fair trade? Next, to take your analysis a little further, work as a whole group to consider the sample analysis that is provided on the reverse side of the handout. This analysis also brings in some critical considerations regarding "fair trade" as an alternative to "free trade." Do you agree with this analysis? Are there points that you would like to add? Are there things you missed?

## EXTENSION ACTIVITY

5. Extend the ethical complexity of the free trade/fair trade debate by introducing critique of the global Fair Trade movement. For instance, it has been argued that Fair Trade USA (original producer of the text of "What is Fair Trade?") is primarily a "marketing model for ethical consumerism," which may not always benefit local farmers directly.[6] The first section of Colleen Haight's article (link provided) is a good starting point for this, but you might also conduct a search for additional

---

6   See Colleen Haight, The problem with Fair Trade coffee, para. 6.

resources. How does this additional perspective impact your thinking through the four questions guiding the activity?

6. The American television series *The West Wing* takes up the complexity of free trade during an episode entitled "Somebody's Going to Emergency, Somebody's Going to Jail" (Season 2, Episode 16). Through a series of scenes in the episode, White House staffer Toby Zeigler – a politico whose strong progressive political positions are checked by his expertise in crafting the most publicly digestible messages – is assigned the difficult task of dialoguing with an unruly group of protesters challenging the World Trade Organization about the White House's policy stance on free trade. To see the full complexity of the screenwriters' arguments about free trade, source the whole episode (on DVD or various streaming platforms) and show all of Toby's scenes to your group. If you cannot access the whole episode, one critical scene in the sequence is available on YouTube.[7] In this scene, Toby exclaims, "free trade stops wars! (… and we figure out a way to fix the rest)." Of course, such complex issues cannot be reduced to slogans. Still, ask your group to wrestle with the good intentions that free trade advocates have for the world. Can we make guesses about Toby's ethical reasoning toward this position?

---

[7] Jonathan Rick, "Toby Ziegler From the West Wing Explains the Benefits of Free Trade."

# HANDOUT: WHAT IS FAIR TRADE?

According to Fair Trade USA's founder, President and CEO Paul Rice, "Fair Trade makes free trade work for the world's poor."

Free trade is the economic theory that the market should be allowed to flow without government intervention. Purists want to get rid of all trade tariffs, subsidies, and protectionist economic policies. However, it is these very regulations that stop commodity prices from fluctuating uncontrollably. This laissez-faire theory aims to reach market equilibrium – where supply meets each demand. What free trade supporters fail to consider is the fact that, sometimes, the means to get that supply is not all that fair.

Historically, free trade has left small-scale producers behind as large subsidized companies start to take over their industries. While large contracted farms can afford to sell commodities at lower prices, local farmers, who have traditionally supplied these products, are driven into debt. The only way these farmers can compete with subsidized farms is to lower their product prices to the point where labour is free and quality of life is unsustainable.

In the case of coffee growers, these producers lack information on the real market value of their commodity, which easily makes them victims to unfair market deals that take advantage of their inexperience. Additionally, these farmers often lack access to credit and are forced to take quick cash from buyers who offer to pay a fraction of what their crop is worth.

Fair trade helps level the playing field by equipping the farmers with tools – information and training – they need to receive fair prices for their products. The Fair Trade system aims to provide greater market access to farmers, which gives them a larger say in how much their product is worth. Fair Trade is seen as "market-based" because it relies on socially-conscious consumers to support the movement by purchasing Fair Trade products. Through their conscious purchases, consumers tell companies that they care about the farmers and workers who produce their products. Fair Trade aims to address the underlying inequities caused by poverty and lack of access to market information that free trade ignores.[8]

---

[8]    Extract from: Fair Trade USA, http://fairtradeusa.org/what-is-fair-trade/faq

# HANDOUT: CONTINUED

**Here we provide, some of our own illustrative reflections on the questions raised in Activity 10.1 to help deepen the discussions:**

### Inconsistencies, contradictions and oppressive relations

One of the key contradictions highlighted in this story revolves around the word "free" and the implication that everyone has equal opportunity to participate and benefit. However, this promise of freedom veils the problem of unequal power relations and unequal access to resources and credit. This makes some members of this "free" system more vulnerable than others and thus makes the system far from fair. Within the free trade economic model there appears an impossibility that freedom and fairness are both attainable, while within the fair trade system both are possible. In the case of free trade, the rich dominate the poor, the poor are subjugated to the whims of the changing market and the poor are exploited in terms of their well-being and their labour and so is the land on which they grow their crops.

### Challenging or supporting the dominant economic system

Fair trade has been proposed as a more equitable trade option, but sometimes there are constraints on change. The powerful who have control of the market may have significant influence at policy level and therefore have vested economic interest in keeping the status quo. Consumers can exert considerable power through their actions as their selection of fair trade or locally bought products can influence the market. Thus the action of buying free trade or local products can be considered to have significant transformative potential. Fair trade is committed to deep structural change in that it is challenging a dominant economic model that threatens the freedom and well-being of the more vulnerable members of society.

### People's relationships with themselves and others, including the more-than-human world

The article above argued that fair trade relies on the actions of conscious consumers to support the movement. This may require not only paying a bit more for products, but it may mean changing your way of eating, your way of dressing, your way of being. Careful consideration of the effect of your buying power on farmers and rural communities; and following through by purchasing such products where possible, shows your care for the farming community and the workers who produce the products.

### Implementation challenges

A potential constraint on fair trade products is that they can be inhibitively expensive when people are being paid the full worth of their product. Another reason that fair trade products become expensive is that these products are not that widespread, and are often imported from other countries. This not only affects their cost, but also has implications for the environment in terms of creating substantial "food miles" for the product and subsequent contribution to global warming through release of fossil fuel.

The questions embedded in Activity 10.1 were inspired by Bhaskar's critical realist dialectic.[9] The four areas of questioning – about inconsistencies, contradictions, and oppressive relations; about emancipatory possibilities that challenge the status quo; about people's relationships with themselves and others; and about the challenges of implementation in context – are designed to help us think about what we need to know and consider in the process of deciding on an ethical action. These questions were used to look critically at the concepts of free and fair trade from a theoretical and abstract perspective. How can these insights help people to decide what to do in their own unique and specific circumstances? Every choice made or action taken will have different challenges and different implications in different contexts.

In order to meet the challenge of making considered, grounded and context-based choices and actions, the next two activities focus on the need to make context-specific decisions. This requires an understanding of the actions taken by particular groups or individuals in particular times, places, and circumstance. Bhaskar calls this an understanding of the concrete singular.[10] For example, both farmers and consumers of different products make choices shaped by particular times, places, and circumstances. They need to consider their own individual needs and how their choices might affect others, the place where they live (what is available where they live?), their socio-economic circumstances (can they afford to make changes and do they have the "luxury" of choice when it comes to certain farming approaches or products to buy?), and more. A careful analysis and choice for an ethical action would require each farmer's choice, each consumer product, and each person's choice regarding that product to be considered for its merit in the light of these particular circumstances.

## LEARNING ACTIVITY 10.2: EXERCISING CHOICE AS PRODUCERS AND CONSUMERS

The next activity sets out to help us, as farmers/producers and/or consumers of certain products, to take ethical action. Here we stress what our own choices and options are. That is, the concrete actions that can follow for each individual.

With these thoughts as background, you can try out and apply some thoughts on free trade and/or fair trade in your own context.

1. Within the group try to find examples of peoples' experience of different producer or consumer choices and actions they have made or

9    Roy Bhaskar, *Dialectic: The pulse of freedom*.
10   Ibid.

have thought about. Can you find examples of producer or consumer actions such as:

- refusing to buy products that are not rooted in fair and/or environmentally friendly practice,
- actively seeking out fair trade products,
- actively seeking markets that are more fair for the product you grow/sell,
- becoming more attentive to product labels or rights and regulations as outlined in policies and constitutions, or
- doing further research into burning questions.

For example, you may have thought about:

- whether it is a good to buy shoes from a company known for its poor working conditions or child labour practices,
- or you may find yourself choosing between a beauty product tested on animals versus a more expensive one that is anti animal cruelty, or
- you may simply find yourself deciding between free-range eggs or eggs from factory farms, or
- between products overly packaged in plastic and polystyrene versus products minimally packaged in recyclable cardboard or paper.

Considering the discussion in the introduction which introduced the idea of actions in support of, or in opposition to, the status quo, as well as the idea of direct or indirect actions:

- Discuss the actions suggested by the participants in your group.
- Ask individuals to explain why they have made these particular choices. As a group, can you see any links to the first activity's guiding questions that consider: inconsistencies, contradictions, and oppressive relations; different possibilities that challenge the status quo; people's identity and relations with others and the more-than-human world; and challenges in particular contexts.

2. Bring two or three items of the same kind of product into the classroom. These could be different coffees, teas, chocolates, sugars, or other products. One of these products should be a fair trade option, another not, and a possible third could be a locally grown and minimally packaged item.

TIP: Local farmers markets, and in-season fruit and vegetables have a higher chance of being grown in your country. Such choices would have the added benefit of reducing food miles, or the distance that food must travel from its production location to where it is consumed.

You can Google "fair trade" "against testing on animals" or "environmentally conscious" or "Eco-friendly" or "organically grown" products in your own country and find something appropriate for your group. What is important is to choose something that IS available to you or some product that might be grown in your local area so that you are familiar with it and are affected by the product in some way or another. An alternative is to bring in photos with prices and clear information from the product label.

In your group, collectively choose one of your preferred products from the samples brought to class and discuss. Why would you choose to buy this particular product in your local shop or supermarket? If you come from a farming area, would you propose that your locally grown products should be fed into a free trade market system, fair trade market system, or simply sold to local shops?

Explain your choices by using the questions introduced in Activity 10.1.

The challenge with fair trade and eco-friendly and health products is that each case is different and has different challenges. Developing ways of being in the world that fight against domination, subjugation, and exploitation can be a set of ground rules for all of us. However, different organisations, communities, and individuals need to find different and do-able ways to challenge these features of oppressive economic systems while embracing the pursuit of freedom and human and environmental health and well-being.

## LEARNING ACTIVITY 10.3: ETHICAL ACTION FOR RHINOS

3. Another way to deepen our understanding of Bhaskar's thinking tools is to see how they might help us to choose ethical actions when presented with a range of existing options. This activity presents three stories (Case studies handouts) that illustrate three concrete actions taken in response to the burgeoning rhino poaching tragedy in South Africa.

   Read each of these and decide:
   - Which, if any, of these case studies are you most drawn to and why?
   - Did you use any of the questions in Activity 10.1 as thinking tools when weighing and judging the options? Which of these tools did you find yourself using?
   - What alternative actions could you propose to deal with the problem of rhino poaching?

# HANDOUT
# CASE STUDY 1: BLACK MAMBAS

## WORLD'S FIRST ALL-FEMALE PATROL PROTECTING SOUTH AFRICA'S RHINOS[11]

THE BATTLE AGAINST the poaching that kills a rhino every seven hours in South Africa has acquired a new weapon: women. The Black Mambas are all young women from local communities, and they patrol inside the Greater Kruger national park unarmed. Billed as the first all-female unit of its kind in the world, they are not just challenging poachers, but the status quo.

The Mambas are the brainchild of Craig Spencer, ecologist and head warden of Balule nature reserve, a private reserve within Kruger that borders hundreds of thousands of impoverished people.

The private reserve's scientists and managers have had to become warriors, employing teams of game guards to protect not only the precious rhinos but lions, giraffes, and many other species targeted by poaching syndicates. The Mambas are their eyes and ears on the ground.

When the poaching crisis started – in 2007 just 13 rhino were killed in South Africa – Spencer ... developed an approach that he says addresses the huge economic and cultural divide between the wealthy reserves and local communities, which he believes drives poaching.

Arrests in Kruger show that the poaching crews are not only foreigners but local South Africans from poor communities. Rhino horn is priced higher than the street price of cocaine and Spencer says cash from poaching turns communities against the park. "The problem really is that there is this perception that has developed in the communities outside the park, they see a uniformed official and think we are the sheriff of Nottingham, they see the poachers as Robin Hood."

"We are not going to police the problem away," he says, standing in the shade of an acacia. "This war will never be won with bullets." In a bid to engage communities outside the park fence, the reserve hired 26 local jobless female high-school graduates, and put them through an intensive tracking and combat training programme. Kitted out in second-hand European military uniforms, paid for by donations, the women were deployed throughout the 40,000 hectare reserve, unarmed but a visible police presence. The numbers suggest the approach works. In the last 10 months the reserve has not lost a rhino, while a neighbouring reserve lost 23. Snare poaching has dropped 90%.

Leitah Michabela has been working as a Black Mamba game guard for the last two years. "Lots of people said, how can you work in the bush when you are a lady? But I can do anything I want." Michabela and the other 26 Mambas are looked up to by the young women in her village as heroes, within the same communities the poachers come from. "I am a lady, I am going to have a baby. I want my baby to see a rhino, that's why I am protecting it."

The reserve uses a team of 29 armed guards, 26 unarmed Black Mambas, and an intelligence team that seeks to stop the poachers before they can kill. The Mambas' main job is to be seen patrolling the fence. They also set up listening posts to hear vehicles, voices and gunshots and patrol the reserve on foot, calling in the armed guards whenever they find something.

Collette Ngobeni sits quietly on top of the landrover with a spotlight under the light of the full moon. She and her team can be seen from miles away, a visual reminder to any poacher of this community's commitment to protect their rhinos. "If we work together as a community we can work this out. People need to open their minds, their hearts. It's not about money, it's about our culture, our future," she says. ∎

---

[11]  Extract from: Jeffre Barbee, "World's first all-female patrol protecting South Africa's rhinos."

# HANDOUT
# CASE STUDY 2: TOENAIL POSTINGS

**Artist posts toenails to embassy in rhino-poaching protest[12]**

*You've got mail, China: toenails.*

A Grahamstown artist – fed up with the escalating rhino poaching numbers – has put his protest in the post. He'd signed the petitions. He'd been to the meetings.

But feeling that his well-meaning outrage was doing no good, Mark Wilby clipped off his toenails, slipped them in an envelope and posted them to the Chinese embassy in Pretoria.

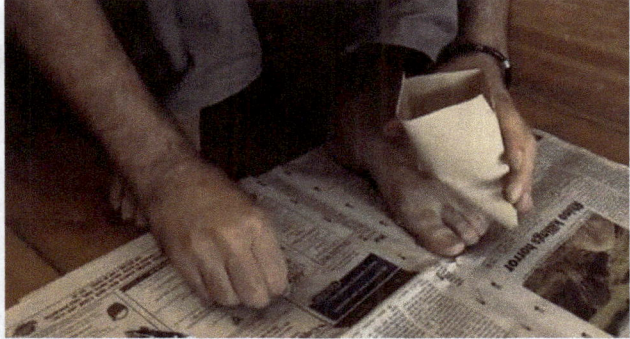

And in a YouTube video of his protest, *sorry China*, he's calling on fellow South Africans to do the same. At least 467 rhinos have been killed so far this year.

*An artist and academic based at Rhodes has come up with a plan to protest against the Chinese governments silence on the issue of rhino poaching. Using Youtube and social networks he has put out a mass appeal to concerned objectors to "nail them" by posting their toe nail clippings to the Chinese embassy. Picture: youtube*

Fuelling the trade is an unscientific belief held in some Asian countries that rhino horn has medicinal qualities. In reality, it's made of keratin – the same inert protein as your hair, fingernails and toenails. "But somewhere in the East, people see it differently and we don't understand that," says Wilby. "It's not enough to say rhino horn is useless. We need to find out more about the market, to understand this thing so we can address the demand."

Wilby said petitions and bumper stickers – outraged South Africans protesting to fellow outraged South Africans – weren't talking to the right people. "I'm not saying the embassy or the Chinese government are complicit," he says. "But we have to persuade the government to help us understand this business."

The embassy in Pretoria has previously stated that the Chinese government shares SA's stance on rhino poaching. They said they had not yet received any nail mail by yesterday afternoon.

Wilby did include a return address, "with trepidation. Sending strange objects to an embassy is not the coolest thing to do."

---

[12] This is an extract from Kristen van Schie's "Artist posts toenails to embassy in rhino-poaching protest."

# HANDOUT
# CASE STUDY 3: INFLAMMATORY DISCOURSE

Figure 10.1     Sign to deter poachers
(Photo courtesy of Anne & Steve Toon Wildlife Photography)

## SOME THOUGHTS ABOUT THEORY

This chapter has been influenced by the work of Roy Bhaskar, especially his book titled *Dialectic: The Pulse of Freedom*.[13] For Bhaskar, the pulse of freedom is a yearning for freedom from errors, deficiencies, omissions, inconsistencies, incompleteness, and contradictions that characterise practices in our lives. Using this idea, we can consider how to think about action as an ethics-led process towards a society and a planet that is freer, more just, and more equitable.

The activities in this chapter and four questions introduced in Activity 10.1 have been influenced by four elements of Bhaskar's critical realist dialectic:

**Element 1:** What is the nature of the world in which we "must act?" There are positive aspects of the world where we see actions of peace, love, solidarity but there are also inconsistencies, injustices, and inequalities. We need to understand the powers of this world – both the powers of oppressive relations that enable and perpetuate these injustices and inequalities, as well as the emancipatory powers that enable the positive aspects of our world.

**Element 2:** By acting in the world we, as agents, have the power to actively work with, or against, or to redirect the status quo. There are many ways we could act, so we need to think about how we EMPLOY the powers of the world and whether we want to use these powers to support the status quo or to look for ways of acting differently in and on the world. Also we need to be conscious of latent powers that are hidden, suppressed, or dormant that perhaps we are not aware of through what we have seen and experienced.

**Element 3:** We need to weigh up all of our possible actions in terms of how they affect ourselves, and our interactions with other people and the more-than-human world.

**Element 4:** There is no one-size-fits-all action that can be imposed on all humanity. Holding our knowledge of the first three points in mind, we then need to agree on what action is right, or morally defensible, for an individual or group of people in a certain place, in a certain time.

In an elaboration of the work done by the activities in this chapter, Figure 10.2 highlights eight tools that have been used to develop these activities and that can be linked to these four elements of Bhaskar's dialectic:

---

[13] Roy Bhaskar, *Dialectic: The pulse of freedom*.

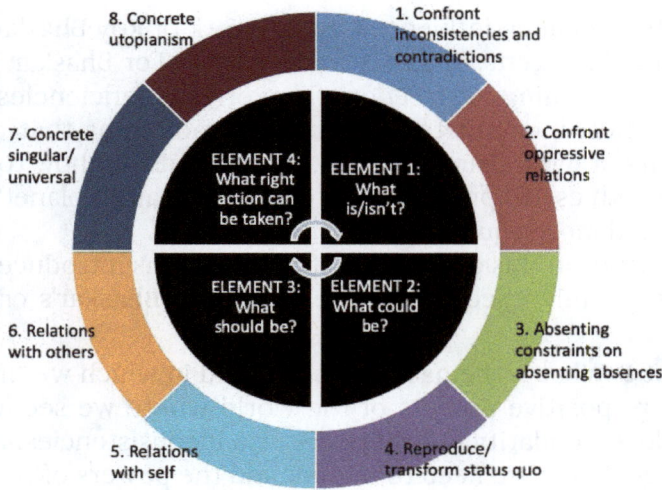

Figure 10.2    Tools for determining ethical action

The eight tools outlined in this figure are described in more detail below.

The first tool used in the preceding activities was the identification of contradictions and inconsistencies in the world. The types of contradictions we might find in our analyses are instances where a system, agent, or structure is pulled in different directions by opposing "rules" or "principles" with different roots, powers, and influences.

For instance, religious, cultural, or constitutional systems might pose conundrums for action. For example, how does a poor farmer, obliged to clear more and more forest to grow crops, abide by trickle-down rules for capital gain – a capitalist "rule"? How does she provide enough money to feed and educate her children – their constitutional right? At the same time how does she obey the rules of managing a healthy ecosystem – a rule which may have been embedded in cultural or Indigenous knowledge, but which is undermined by the other rules posed here?

There could also be internal contradictions whereby a system, agent, or structure undermines itself in some way. For example, a concern with human health has led to the widespread use of DDT (Dichlorodiphenyltrichloroethane) to control the spread of mosquitoes in Africa, yet the persistent nature of this chemical leads to different threats to human and ecosystem health.

These examples illustrate how we can start to look for the essence of conundrums that we are faced with. Sometimes we are overwhelmed by the complexity of things that we instinctively know are "not right", but we cannot see where to begin to tackle the problem.  This tool and the following tool can be seen as a place to start to "get the ball rolling" in our quest for social change.

A second tool of Bhaskar's that helps in beginning to chip away at what we instinctively know is not right, is that of looking at oppressive power relations. Bhaskar distinguishes oppressive relations that enable "capacity to get one's way against either (i) the overt wishes and/or (ii) the real interests of others."[14] Such relations have "negative characteristics such as domination, subjugation, exploitation and control" by forcing someone to do something they would not otherwise do, by narrowing their choices or by socialising someone into a worldview that is ultimately against their own interests.[15]

For example, in South Africa such oppressive relations were evident in the Apartheid curriculum for black children that emphasised gardening and agriculture in schools.[16] Its recipients resisted this curriculum, as it was associated with a narrowing of "career" choice and a view of the role of black people as labourers on farms or gardeners in the homes of the colonial rulers. In another example, a contemporary and popular environmental learning activity – that of making something useful from waste – could be critiqued if done in such a way that it teaches those from poor social-economic backgrounds to "live off the waste of others". It is such relations of domination, subjugation, and control; that, if targeted for change action, can lead to a resolution of contradictions and inconsistencies.

A third tool helps us decide what to do in order to change – or absent – the conditions we identified through the first two tools, or to change any conditions that prevent us making the desired change. Bhaskar refers to this as "the process of absenting constraints on absenting absences (ills, constraints, untruths, etc.)."[17]

Let us look at "upcycling" projects as creative and quirky, and a way of creating employment for those in desperate need of a way of making a living. For example, weaving mats from discarded plastic bags can be a colourful feature or talking point for a bathroom floor while providing an income for someone in need. It is a step in the right direction for raising awareness and for reducing waste. Or is it? We can ask – what awareness are we raising really? Are we constraining the absenting of waste by "glorifying" waste, or by "smoothing the conscience" of rampant consumerism? This is while the amount of plastic by which we reduced the plastic gyre in the Pacific Ocean – when we bought a plastic mat – was minimal to the point of insignificance. Did you know:

---

14  Roy Bhaskar, Dialectic: The pulse of freedom, 153
15  Mervyn Hartwig, Dictionary of critical realism.
16  Nelson Mandela Foundation. Emerging voices.
17  Roy Bhaskar, Dialectic: The pulse of freedom, 197.

The vast majority of all the plastic ever produced still exists (about 4.9 billion tonnes of it) ... A third of all plastics end up in fragile ecosystems like the oceans, 40% is buried in landfill and, of the remainder, most is burnt whilst only 5% globally is recycled.[18]

So, if our two problems are the absence of a healthy environment and unemployment, we are arguing that by embracing and celebrating plastic and by hiding the magnitude of the problem, we are constraining the absenting of these absences. What other ways can we think of for absenting these problems? Ellen MacArthur says:

Looking beyond the current take-make-dispose extractive industrial model, a circular economy aims to redefine growth, focusing on positive society-wide benefits. It entails gradually decoupling economic activity from the consumption of finite resources, and designing waste out of the system.[19]

The fourth tool helps us think about how the nature of our actions – or actions through non-action – could have varying effect that could either:

- be reproductive  – consistent with the status quo – or transformative – aimed at social change – and/or
- involve indirect action – for example not to buy consumer goods that have a particularly high environmental footprint or to participate in the development of policy that legislates the boundaries of agential action – or direct action – for example to consciously buy organic or "green" products or to make and/or sell one's own alternative products – and/or
- be attentive to research life cycles and actively read labels on consumer products, or to be inattentive and buy goods without paying attention to the life cycle of the product and its ingredients.

The fifth and sixth tools are influenced by what Bhaskar refers to as the moment of totality that includes "transactions with self and transactions with others" respectively. The latter "transactions with others" includes transactions with non-humans. Price argues that "if we are part of the world, then to take care of oneself, is to take care of the world, but to sacrifice oneself in certain circumstances to ensure the well-being of others is still to take care of oneself, since the self is part of that totality."[20] A similar view of the relationship between humans and nature is evident in the work of Plumwood who argues that we will "need to reconceive the human self in more mutualistic terms, as a self-in-relationship with nature, formed not in the drive for mastery and control of the other but in a balance of mutual transformation and negotiation."[21] This mutualistic relationship or "totality" implies that we do not need to set human

---

[18]  Henry Boucher, "Fantastic plastic - too much of a good thing?"
[19]  Ellen MacArthur Foundation, "What is a circular economy?"
[20]  Leigh Price, Transdisciplinary explanatory critique of environmental education251.
[21]  Val Plumwood, "Paths beyond human-centeredness," 101.

concerns and nature concerns in opposition to each other; you can be doing something that may be altruistic, but still receive a benefit at the same time. That does not nullify the altruistic impulse.

Also, beyond the non-altruistic self-interest implied in Plumwood's discussion, Price explains that assuming "humans and non-humans are a 'totality', connected in an intimate web of relating, then awareness of this connection, or sympathy, leads to moral sentiments which may be altruistic."[22] Elaborating on this notion of altruism, Bhaskar and Parker argue that **care**:

> can form the basis for a new immanent humanist ethic that is capable of sustaining a continuing commitment to human emancipation and self-realization, rather than just mere survival.[23]

The seventh tool can help individuals or groups to decide on an action and a way forward. The tool highlights agency and ethical practice. This tool is aimed at context-sensitivity where "all individuals, though sharing a common humanity, [are treated] as ethically different."[24] This means that different actions would be relevant for different places, times, and individuals so that rights and freedoms are "historically specific, concretely singularised and open; the norms enshrining them will be the norms of the communities concerned, not those of the abstract universal."[25] These quotes describe Bhaskar's notion of the "concrete singular", as in tool seven. Bhaskar explains "concreteness" further as: "its nearest synonym might be 'well-rounded', in the sense of balanced, appropriate and complete for the purposes at hand."[26]

The notion of concreteness is also important in the eighth and final tool which entails the construction of "models of alternative ways of living on the basis of some assumed set of resources",[27] and which requires us to consider what we have and how we can realistically work with it so that we are not misleading ourselves with idealistic propositions. That is considering actions that are not based on an idealistic Utopian vision but on a notion of "concrete utopianism" – a notion that acknowledges the "role of creative fantasy ... that yields at once hope and possibility."[28]

One way to look at the complexity of the diagram above, its four elements and eight tools is that it enables us to think about ethical action from four perspectives:

- What **is and what is "not right"** (Element 1)?
- What **could be** different (Element 2)?

---

22  Leigh Price, Transdisciplinary explanatory critique of environmental education, 250.
23  Roy Bhaskar, & Jenneth Parker, vii.
24  Mervyn Hartwig, *Dictionary of critical realism*, 74.
25  Ibid., 492.
26  Roy Bhaskar, *Dialectic: The pulse of freedom*, 1993: 119.
27  Ibid., 395.
28  Ibid., 209.

- What **should be,** considering a holistic vision of self and others (Element 3)
- What realistic, concrete **action can we take** (Element 4)?

## References

Barbee, Jeffre. "World's first all-female patrol protecting South Africa's rhinos." *The Guardian*, February 26, 2015, accessed February 12, 2021, https://www.theguardian.com/environment/2015/feb/26/worlds-first -all-female-patrol-protecting-south-africas-rhinos

Bhaskar, Roy. *Dialectic: The pulse of freedom*. London: Verso, 1993.

Bhaskar, Roy. *Possibility of naturalism* (3rd edition). London: Routledge, 1998.

Bhaskar, Roy. *From science to emancipation: Alienation and enlightenment*. New Delhi: Sage, 2002.

Bhaskar, Roy, and Parker, Jenneth. "Introduction," in *Interdisciplinarity and climate change: Transforming knowledge and practice for our global future*, eds. Roy Bhaskar, Cheryl Frank, Karl George Høyer, Petter Naess, & Jenneth Parker. London: Routledge, 2010, vii-xiii. https://doi.org/10 .4324/9780203855317

Boucher, Henry. "Fantastic plastic - too much of a good thing?" *Sarasin & Partners*, November 27, 2018, accessed June, 12, 2019, https://www .sarasinandpartners.com/global-home/insights/article/article-460 -nov18-fantastic-plastic---too-much-of-a-good-thing

Democracy Now. "Liberian Nobel Peace Prize laureate Leymah Gbowee: How a sex strike propelled men to refuse war," April 27, 2015, accessed November, 25, 2015, http://www.democracynow.org/2015/4/27 /liberian_nobel_peace_prize_laureate_leymah

Ellen MacArthur Foundation. "What is a circular economy? A framework for an economy that is restorative and regenerative by design," 2017, accessed July, 11, 2019, https://www.ellenmacarthurfoundation.org /circular-economy/concept

Haight, Colleen. "The problem with Fair Trade coffee." *Stanford Social Innovation Review*. Summer, 2011, accessed January 15, 2019, https://ssir .org/articles/entry/the_problem_with_fair_trade_coffee

Hartwig, Mervyn. *Dictionary of critical realism*. London: Routledge, 2007.

Nelson Mandela Foundation. *Emerging voices: A Report on education in South African rural communities*. Cape Town: HSRC Press, 2005.

Plumwood, Val. "Paths beyond human-centeredness: Lessons from liberation struggles." In *An invitation to environmental philosophy*, ed. Anthony Weston. New York: Oxford University Press, 1999, 69-105.

Price, Leigh A. Transdisciplinary explanatory critique of environmental education: Business and industry. Unpublished Doctor of Philosophy study, Rhodes University, Grahamstown, 2007.

Rick, Jonathan. "Toby Ziegler From the West Wing Explains the Benefits of Free Trade." September 5, 2016, *YouTube* video, 0:55, accessed January 15, 2019, https://www.youtube.com/watch?v=8dGkiJcEK78

van Schie, Kristen. "Artist posts toenails to embassy in rhino-poaching protest," *The Star*, October 24, 2012, accessed February 12, 2021, https://www.iol.co.za/the-star/artist-posts-toenails-to-embassy-in-rhino-poaching-protest-1409607

# CHAPTER 11

## IMPLICATIONS FOR TEACHING AND LEARNING

**Pedagogical intent:** This chapter presents an analytical tool that we are calling a heuristic. This heuristic is designed to help educators reflect on their own conceptions of ethics, to help make decisions about content worthy of inclusion in their lessons and to examine the ways that this content might be educationally presented. The Earth Charter, as a frequently referenced public document and statement of principles, is used as an example to illustrate the application of this tool. This application also illustrates some of the complexities and challenges that are inherent in educational activities. In the end, this heuristic is intended for use in critiquing current discourses, evaluating new initiatives, and finding one's own educational place within present debates.

**Activities:**                   11.1: Conceptualising ethics

                                   11.2: Mapping a heuristic

                                   11.3: Mapping the Earth Charter

                                   11.4: Exploring the tensions

                                   11.5: Extending the heuristic

**Main theoretical resources:** John Ralston Saul, who challenges us to evaluate and resist authoritarians within our cultures; Aldo Leopold and Anthony Weston who, in their own ways, suggest that environmental ethics is in need of a great deal of exploration; as well as Karen Warren, Val Plumwood, and Arne Næss who all push us to see ethics as existing beyond simple logic and rationality.

## Ethics: Exploring "what do you mean?"

The world is getting extremely hot; species are rapidly going extinct; biodiversity is diminishing; the gap between rich and poor is increasing. Humans appear to be on a suicide mission. Many folks who should know better, are oblivious to mounting evidence. To make matters worse, the very language of environmental change is being co-opted. For example, "sustainability" is often just used, amongst other things, to sell products. Illustrating this point is an advertisement from a leading Canadian newspaper that used the slogan, "sustainable excitement" in a pitch to sell Volkswagen's automobile, the Jetta.[1] Of course, this is not a value-neutral statement. Implicit in this slogan is a position that suggests the business of

---

[1]   Globe and Mail, Advertisement, April 1, 2013, A14.

"sustaining" car sales trumps concerns about environmental sustainability. Here the plasticity of the term "sustainability" can sustain and reinforce values and attitudes that are harbingers of ecological catastrophe. Seen this way, tacit expressions of value can be seen everywhere.

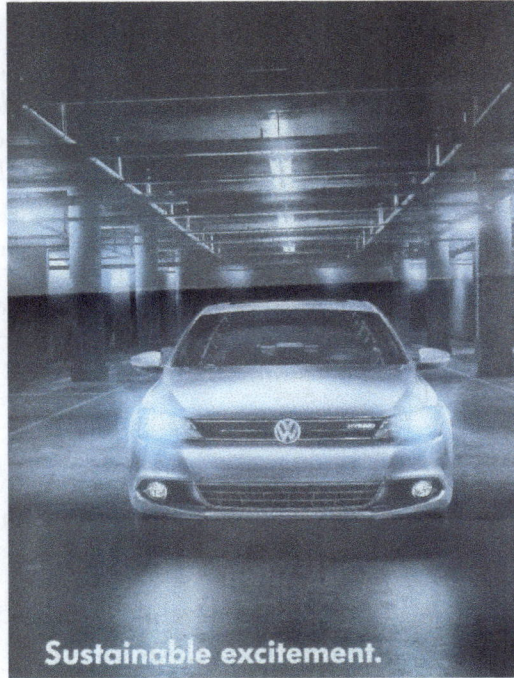

Figure 11.1    Sustaining car sales; but is that sustainability? (Photo by author, Bob Jickling)[2]

A worse co-option was observed at the Copenhagen climate summit (UN COP 15). Evidence suggests that discourse was shifted subtly from the language of decarbonisation to the language of adaptation.[3] While adaptation is important, it is dangerous as a replacement for reducing carbon emissions. These observations underscore the trouble we are in.

This does not, however, take away from the importance of also having to work on adaptation. Climate change is real and the Earth is heating up. We must concern ourselves with this reality and the adaptation that it demands. However, we must also be critical and seek to understand the assumptions and values buried deeply in the global conversations about climate change. This process of critically teasing out the tacit values in cultural artefacts, and then debating their merits, is a form of ethics.

---

[2]    This image was taken from an advertisement that ran in *The Globe and Mail*, April 1, 2013, p. A14.

[3]    See for example: Ingolfur Blühdorn, "Political sociology and the cultural framing"; Bob Jickling, "Normalizing catastrophe", 161-176.

This critical work seems to be paying off. The IPCC reports from 2014, and the most recent COP21 agreements, propose a form of climate resilient development or responsiveness that encompasses both adaptation *and* mitigation through decarbonization. Still, climate activists suggest that this might not be a fully adequate response, and propose a framework for human development that involves dematerialisation, decoupling of innovation from resource extraction, making 100% commitments to renewable energy, and addressing fundamental social injustices which involves nothing less than a fundamental re-orientation of economic paradigms. Thus, critical work continues.

Meaningful change does not come from continuously doing the same thing. More science will be helpful in keeping issues like climate change current. It will set up a meaningful context for change. Still, something more radical may be essential. Arne Næss once said, "we have had for a long time more than enough ecological knowledge about how to mend our ways."[4] For him, the important question was around value priorities. We agree. We also suggest, throughout this book, that these value considerations need to be coupled with educational actions that can enable reconstructive actions and transformative practice. This, too, can be framed as ethics in action.

In times of change, uncertainty, and stress, interest in values and ethics increases, as Andrew Sayer[5] reminds us. Unfortunately, ethics are messy and uncomfortable. The term has multiple meanings, and few educators have had much preparation in ethics.

The examples used in these introductory notes begin to highlight both the messiness of ethics, and a broad range of possibilities for ethical engagement – in our own lives and as educators. Our first activity is intended to explore this range of possibilities a little further. As authors, we have noticed that most people have an initial response to what ethics are, and what they mean when they use the term "ethics". Often within a classroom or workshop-sized group, ethics can be expressed across a range of possibilities. The objectives of the first activity are to make visible the diversity meanings that can occur with a group, and then to use these observations to open a discussion about this range of possibilities. This is a prelude to a more in-depth look at how educators may position themselves in professional settings.

## LEARNING ACTIVITY 11.1: CONCEPTUALISING ETHICS

1.  So, what do you think of when you hear the word ethics?

---

4   Arne Næss and Bob Jickling, "Deep ecology and education," 55.
5   Andrew Sayer, *Realism and social science*.

Take a moment to write down, on a post-it or small piece of paper, what your own conception of ethics is? Share a few examples of these conceptions with the rest of your group. Save this note, as we will use it later.

## Teaching, learning, and ethics

Our abilities to be innovative are inextricably linked to how we conceptualise teaching, learning, and ethics, and how well these conceptions work together. To help explore the possibilities, we use a heuristic developed by Bob Jickling and Arjen Wals[6] for looking at teaching and learning (See Figure 11.2).

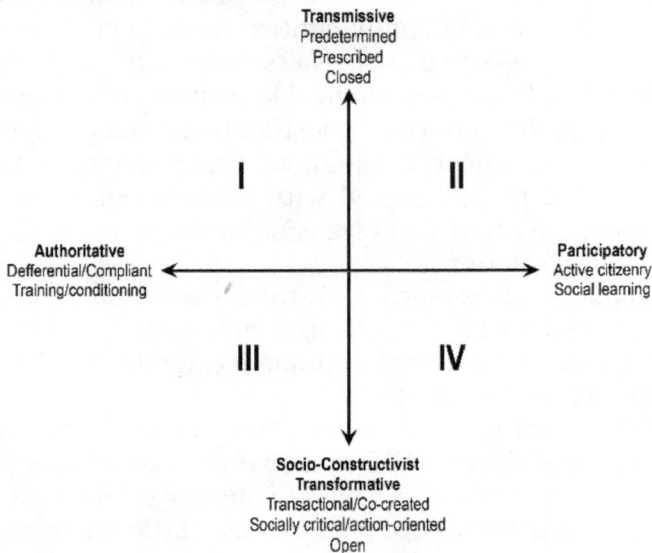

**Transmissive**
Predetermined
Prescribed
Closed

I    II

**Authoritative**
Defferential/Compliant
Training/conditioning

**Participatory**
Active citizenry
Social learning

III    IV

**Socio-Constructivist**
**Transformative**
Transactional/Co-created
Socially critical/action-oriented
Open

Figure 11.2    A heuristic for teaching and learning about environmental ethics using intersecting continua: Intention (transmissive/transformative) and approach (authoritative/participatory) (Figure courtesy of author, Bob Jickling)

As this diagram suggests, one way of framing contesting views is along a pair of continua. The vertical continuum depicts a range of educational and pedagogical possibilities from transmissive on one end to socio-constructivist at the other. Transmissive teaching and learning is essentially about the transmission of facts, skills, and values to learners.

Socio-constructivist and transformative education and/or ethics, on the other hand, reflect a belief that education and ethics are emergent achievements. Here knowledge and understanding are co-created within

---

6    Bob Jickling and Arjen E.J. Wals, "Globalization and environmental education," 1-21; Bob Jickling and Arjen E.J. Wals, "Probing normative research in environmental education," 74-86.

a social context. New learning is shaped by prior knowledge, experience, and diverging cultural perspectives. As such, it can be transformative. This characterisation provides space for some measure of autonomy and self-determination on the part of the learner. Along the horizontal continuum we can ask whether education and or ethics is mostly about social reproduction at one end, or more about enabling socially active citizenry at the other. Such distinctions can reflect the ways that practitioners imagine citizens interacting within society, and its democratic and/or social practices.

If social reproduction were expected, then "educated" or "ethically considerate" citizens would be expected to work efficiently within existing social frameworks. They would primarily be deferential and compliant in taking their place within hierarchical social structures and power relationships.

On the other hand, if social engagement is expected, then "educated and/or ethically considerate citizens" would actively participate in ongoing decision-making processes within their communities. Here, education and ethics are crucial in realising a sense of self, a sense of other, and a sense of community. This position also creates space for self-determination, as individuals and/or members of groups, and greater degrees of autonomous thinking in a social context.

When putting these possibilities together, readers can certainly generate interesting ways to consider educational qualities and conceptions of ethics that might characterise four positions within each of the quadrants defined by the two continua. As a starting point, Quadrant I can represent a place where folks who tend towards authoritative understandings of the world lean towards transmissive educational strategies to implement educational and ethical aims.

Quadrant II can represent folks who employ participatory learning strategies but maintain expectations that preferred knowledge will still be transmitted, and that ethics will still be circumscribed by particular aims. Similarly, educators in Quadrant III may employ constructivist educational strategies while maintaining an authoritative outlook. Learners can work at constructing meaning for themselves, but in a climate that exudes authoritative expectations.

Quadrant IV educators have expectations that learners will be active participants in their education and in this process will construct understandings with high degrees of freedom and autonomy. This is seen as a route to discovering new ways of understanding and being in ethical relationships with the world. Educators aspire to enabling transformative experiences.

Exploring concepts like education and ethics is complex and a two-axes-heuristic is not sufficient to capture the entire shape and scope of the enterprise. Moreover, the positions that are characterised by spaces in each quadrant are clearly idealised. No educational activities are really this tidy, and ethics can be a much more complicated activity, often with internal tensions. It is important, therefore, that we call it a heuristic, rather than a framework. As a heuristic it is meant to be generative – to engage people with tensions within environmental education and ethics, and to challenge them to carefully consider their practices. Also to engage people with the tensions inherent in the heuristic itself – to redesign it, to use alternative axes, or to use it as a stepping stone to construct an altogether alternative analytical tool.

For the time being, use it to begin analysing the kind of a teacher you are. Is your teaching about transmitting knowledge, or is it about constructing knowledge in social learning settings? Do you see yourself, and your curricula as authoritative, or do you see students as participants in co-creating learning opportunities?

To be fair, we probably use all quadrants in our teaching at different times. However, questions remain: What tensions within our own teaching can the heuristic reveal? Where do we want to go in our teaching?

These questions will be explored in the next activity and again later when using, as an example, the Earth Charter as a vehicle for learning.

## LEARNING ACTIVITY 11.2: MAPPING A HEURISTIC

1.  Prepare a drawing of the heuristic described in Figure 11.2, on chart paper, or a black or white board. Make sure it is large enough to be visible to your entire group. If need be, you can make two or three of these drawings and place them around the room.

    Now ask participants to take their statement about ethics (from learning Activity 11.1 above) – written on a post-it with a sticky edge – and place it on this drawing of the heuristic at a spot that seems to best represent their alignment between teaching and their stated view of ethics. Following these placements, encourage participants to survey the range of ethics conceived and expressed by their group. What are the relative merits of these positions? What do they suggest about individuals' teaching strategies? Conceptions of education?

    The primary intention for this activity is to reveal a range of understandings of ethics that typically occur within group settings, and that often mirror a community at large. In our experience, people's thoughts about ethics can range from codes of practice or social rules, to something akin to religion or dogma, to a process of inquiry and reflection. Some see ethics as rational and analytical while others tie it to more emotional qualities such as love, empathy, and respect. Yet, others align ethics not with what a person might think or say, but rather with what they do. In more educational terms, ethics can sometimes be seen as an imposition of someone's own values upon other people's kids.

The point here is to show that there are lots of legitimate ways of talking about ethics and, in doing so, to realise that there can be lots of starting points for a journey into ethics. It can also begin to open discussion about how ethics might be introduced in ever-more educational ways into learning settings.

There can be a couple of things to watch for, however. First, while there is generally a natural range within a group's conceptions of ethics, this is not always so. Sometimes most of the statements about ethics are all placed in the same quadrant. It can be helpful to have some probing questions that help the group to begin thinking outside of a singular box. Our own experiences shared above might be helpful in developing a few probes. Second, people will sometimes place their statements in a place where they want to be rather than where they described in the first phase of the activity. This result need not be a problem and can be used to fuel the discussions suggested above.

## The Earth Charter

An interesting way to test our individual teaching strategies and ideas about education is to see how we might feel about them when put to the test. With this in mind, we suggest trying an experiment. In this case we propose taking a critical examination of the Earth Charter, an international framework designed to guide ethical actions (see handouts). As such, it exists as an interesting cultural artefact.

At face value, the Earth Charter website states that it is, "a declaration of fundamental ethical principles for building a just, sustainable and peaceful global society in the 21st century."[7] As such, it is an initiative that promises to challenge "our" values and to choose a better way. The Charter was conceived as a civil society initiative by Maurice Strong (Secretary-General of the Rio Summit in 1992) and Mikhail Gorbachev (former leader of the USSR). As such, it is the product of an extensive, decade-long, consultation that involved worldwide and cross-cultural dialogues about common goals and shared values. The final text was approved at a meeting of the Earth Charter Commission at the UNESCO headquarters in March 2000. It is currently available on the worldwide web in 52 languages. The Commission was co-chaired by Strong and Gorbachev and consisted of a diverse group of 23 eminent persons from all the major regions of the world. Steven Rockefeller chaired the international drafting committee. The dissemination, adoption, use, and implementation of the Earth Charter is managed by the Earth Charter International Secretariat located at the University for Peace in San José, Costa Rica.

---

[7]    Earth Charter Initiative, "The Earth Charter initiative."

The Earth Charter holds appeal for many educators because it explicitly links the caring for the earth and caring for people as two dimensions of the same task, as stated in the preamble:

> We must join together to bring forth a sustainable global society founded on respect for nature, universal human rights, economic justice, and a culture of peace. Towards this end, it is imperative that we, the peoples of Earth, declare our responsibility to one another, to the greater community of life, and to future generations.[8]

For some, this ambitious work of socio-ecological synthesis gives this document a vision that invites environmental educators to re-examine the significance of their work in fundamental, and new ways.[9]

Other educators are less sure. We recall one African scholar who, when presented a "pitch" for the Earth Charter, commented: "it sounded like another salvation narrative."[10] Ursula van Harmelen elaborates on this concern,

> To impose any ethical framework, no matter how "good" it may be, without subjecting it to constant critical scrutiny and challenge is a denial of human freedom to make informed choices. The imposition of an ethical framework as a given is to reify not only that framework, but to mask possible interpretations of that framework that may in fact be corruptions of the original ideals. Thus, if we were to accept the Earth Charter as a code of ethics without critical and informed analysis this would seem to me to be a form of indoctrination.[11]

She later adds,

> By all means let us use the ethical framework of the Earth Charter in our educational endeavours. But let it be one set of ethics that should be explored and examined alongside other sets, so that if individuals do subscribe to this particular ethical code they have done so based on an informed and reasoned choice, so that they understand their choice, its implications and its consequences.[12]

It could, for example, be said that educational programmes can use the Earth Charter as a critical referent, rather than a set of guiding principles, or universal truths.

So, what are we to do with it?

---

[8] Ibid., 1.
[9] See for example David Gruenewald, "A Foucauldian analysis of environmental education," 71-107; Sean Blenkinsop and Chris Beeman, 69-87.
[10] Ursula van Harmelen, "Education, ethics, and values," 124-128.
[11] Ibid., 125.
[12] Ibid., 125-126.

## The heuristic at work

Whether the Earth Charter is seen as an introduction to the topic of ethics, an appropriate aim for ethics, or a critical referent in a process of exploration of ethical possibilities depends on how ethics and education are conceptualised. Drawing from the heuristic described in Figure 11.2, we have elaborated the analytical capabilities in Figure 11.3. Here we have deployed the original heuristic in the analysis of the Earth Charter as it might be seen when positioned between two teaching force fields. See Figure 11.3.

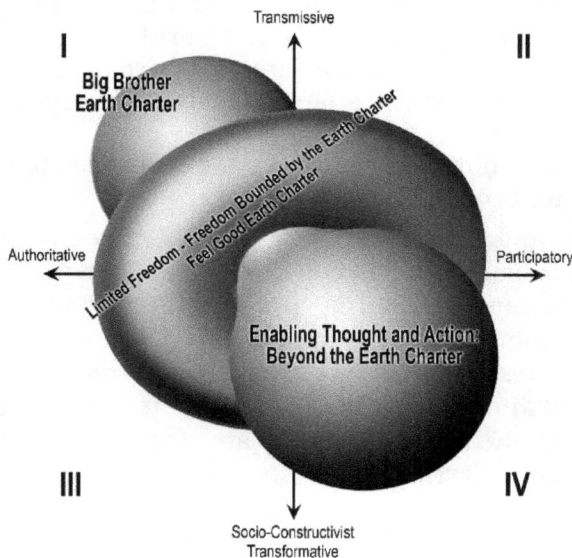

Figure 11.3   Positioning The Earth Charter within two force fields for teaching and learning about environmental ethics (Figure courtesy of author, Bob Jickling)

Quadrant I, seen as *Big Brother Earth Charter*, reflects a prescriptive interpretation and implementation. Here ethics are seen as a transmitted message – these are the principles you should follow – for a compliant citizenry:

■ Ethics are framed as a code that describes appropriate values and expected conduct – a product to implement.

■ The Earth Charter vision and goal statements nod in this direction.

Quadrants II & III are described as *Limited freedom – freedom bounded by the Earth Charter*. Some supporters have responded to concerns raised above. For them educators "must also encourage students to engage actively and critically with the text, rather than to proselytize its values."[13]

They also claim:

The Earth Charter is a normative statement of shared values. However, we see the text not as a programmatic or totalizing structure for teaching about sustainability, but as a plurality of ideas about sustainability broadly conceived.[14]

However, despite encouragements to engage critically with the Earth Charter, to use it as a starting point, to apply it in many ways through a diverse range of settings, and not to proselytise:

- The freedom to explore the Charter's inherent plurality and teach with it in many ways is still framed by the Charter itself.
- So, while students are encouraged to be critical there is a strong authoritative statement that declares the normative eminence of the Earth Charter. This can effectively limit the scope of critical analysis.
- It may feel good, but educators are not encouraged to think outside of the "Charter box".

Thus, In Quadrants II and III there are overtures to participatory engagement and social learning: however, these activities are weighed down by transmissive teaching predilections and authoritative expectations, sometimes implicitly imposed.

Quadrant IV is described as *Enabling thought and action – through and beyond the Earth Charter*. There is ample literature to suggest that environmental ethics can meaningfully and fruitfully be conceived as something beyond the Earth Charter, something other than a code. This quadrant:

- challenges educators to think about teaching strategies,
- challenges educators to go more deeply into possibilities for conceiving of and teaching ethics, and
- portrays the Earth Charter as an exciting document and tool for exploring ethics, but not an endpoint in the on-going evolution of ethical possibilities.

## LEARNING ACTIVITY 11.3: MAPPING THE EARTH CHARTER

1. For this activity, reproduce the heuristic presented in Figure 11.3 and position it in your learning space, alongside the heuristic drawn in Figure 11.2 above. Alternatively, add the analytical descriptions of the various Earth Charter interpretations to the original drawings.

---

[13] Joseph Weakland and Peter Corcoran "The Earth Charter in higher education," 154.
[14] Ibid., 155.

Now have participants take a plain post-it and place it on the heuristic at a spot that represents a place where participants see themselves going in their explorations of teaching learning and ethics. Does the overall "map" of fresh post-its suggest an evolving sense of what ethics can be? Why or why not?

2. Working together, come up with some good stories from your region that represent activities in Quadrant IV. How has the Earth Charter been used to creatively and imaginatively challenge teaching strategies? To encourage the on-going evolution of ethical possibilities?

## Tensions within a working heuristic

It is important to return to the reality of the working educator. As we have acknowledged, all of us who engage with learners find ourselves located in multiple quadrants at different moments in time. Sometimes, we feel that some prerequisites are required before the more critical work can begin. Sometimes time and/or curriculum constraints impinge on our abilities to exercise the educational discretion we feel is needed. These points are just the beginning of what could be a long list.

However, we believe that it is still interesting and fruitful to look at the tensions that can arise between where educators place themselves in the heuristic and where they would ultimately like to be with their teaching activities.

Mirian Vilela[15] is the Executive Director of the Earth Charter International Secretariat at the University for Peace in Costa Rica. Interestingly for us, she extensively uses the heuristic presented here in her classes and workshops. It is important for her that the Earth Charter has the capacity to do new work, that it is not taken for granted, and that it does not simply become empty rhetoric. It is important for her to challenge learners that come to her institution to see the Earth Charter as a living, challenging, and evolving articulation of values and principles. She expects these learners to be critical.

While taking a critical position, Vilela has also identified the importance of work in all areas of the heuristic depending on the stage of your work within a certain group and context. Speaking of the four quadrants, she said "that we really cannot leave the 'rooms' empty; we are not always, all the time in quadrant IV, even if we would like, because it depends on the stage of the group we are working with." "It can", she went on to say, "take time to deepen understanding of the concepts before going on. And work in Quadrant I, for example can be essential."

Some of Vilela's students described their experiences with the heuristic as being like a playground slide. Speaking particularly of Quadrant I, she

---

15 This section is based on a personal communication with Mirian Vilela on March 3, 2017.

said her students indicated that sometimes "we first climb the ladder as they grapple with theory and concepts, then slide down, whizzing towards Quadrant IV. Then they climb back up and slide down again." Many of these students felt that they would like to work more within the fourth quadrant and find themselves often sliding up to the edge of this quadrant. They recognise that this process of climbing and sliding helps them and their students interpret the theoretical possibilities, deepen their understanding, and prepare them to guide the kinds of creative processes that would occur in Quadrant IV. It is easy to imagine that the same kind of process could be at play with educators who find themselves beginning in Quadrants II and III.

Vilela also evokes that the Earth Charter, itself, suggested that learners often need something to work with. "It is", she said "very difficult to develop something like a new ethical position from nowhere or a blank piece of paper. You can start with something, a situation, a context, or a text such as the Earth Charter and build on it." For her, the Earth Charter is not just something to be internalised and applied. Rather, the important challenge is to enable it to do interesting, challenging, and creative work.

Do not simply adopt the Earth Charter, rather "Make it do work!" she concluded.

## LEARNING ACTIVITY 11.4: EXPLORING THE TENSIONS

1. An initial characterisation of educational qualities and conceptions of ethics that fall within each of the quadrants defined by the two continua is provided in the text above. Take some time to discuss, elaborate, or perhaps correct these characterisations? Where did you locate yourself and what good work do you feel that you can accomplish at this location? Would you like to move your teaching and conception of ethics to another location – subtly or significantly? What tensions do you feel between where you are and where you would like to go? How might you overcome the tensions and obstacles?

## LEARNING ACTIVITY 11.5: EXTENDING THE HEURISTIC

1. Using the analysis of the Earth Charter as a backdrop, consider another ethics framework discussed in this book, such as the animal liberation and animal rights theorising in Chapter 4, and imagine how you would like to teach about these ideas. This is an interesting choice, partly due to the more radical tactics of some animal welfare followers who can create an oversimplified image of the ethical underpinnings of their movements. Still, though these theories are flawed when applied to

environmental ethics, are they still worth including in your syllabus? Can the underlying thinking behind animal liberation and animal rights still do useful work? What would you do? How would you expect the learner to engage with animal rights issues?

Deep Ecology might also be a good concept to explore. It seemed to wane in influence with the rise in the popularity of sustainable development, especially after the Earth Summit in 1992. Some thought it was naïve, and idealistic. Given the inability of sustainable development to gain the traction required for change, it seems like a good time to take a second look. Perhaps employing the heuristic can reveal ways to work with the ideas of Deep Ecology in educational settings? Finally, we expect that pedagogical implications for the ideas in each chapter in this book could be usefully tested against the heuristic.

2. As developed in Chapter 4, ethical quandaries are a popular educational activity. Adapt the heuristic in a way that will make it helpful in working out how you would like to present ethical quandaries. Begin with the "sustainable excitement" example presented at the beginning of this chapter. How would you choose to use this material in an educational setting? Also, consider examples presented in Chapter 4. Or find new ones from your own context.

## SOME THOUGHTS ABOUT THEORY

According to the analysis presented here, educational work or environmental ethics located in Quadrant I and/or Quadrants II and III of Jickling and Wal's heuristic could reflect soft pretensions to participatory democracy and socio-constructed understanding and transformation. That is, pedagogical activities may look participatory or socially constructivist but are weighted down by an educator's overriding need for control over outcomes. This need can sometimes be expressed in overt ways, or it can find expression implicitly through teacher actions or subtle steering of subject material or conversation. With expectations for overall agreement about a particular ideology or statement of principles, and no meaningful opportunity to dissent, power slips from citizens (in spite of civil society initiatives) to those that promote these agendas. Furthermore, ideologues are unshakable in the belief that they are on the trail to truth – and to the solution to our problems.

This would come as no surprise to John Ralston Saul[16] who claims that authoritarians, and their courtiers, are fond of order and contemptuous of legitimate doubters. Put another way, this contempt for legitimate doubters can be seen as a way to normalise a particular ideological orientation and undermine democracy.

---

[16] See John Ralston Saul. *The unconscious civilization.*

The critique in this chapter uses the Earth Charter as an example of a statement of principles that could be interpreted as a code of ethics to be implemented. At times, some observers have described it in these terms. Practitioners need to be vigilant. If environmental education is conceived as an invitation to consider what is currently missing in education and ethics, then there are many generative research possibilities. We have also shown that the Earth Charter can be used as a critical referent and a good starting place to begin the challenging and creative work of environmental ethics in educational settings.

As educators we are most concerned about tendencies towards obedience – acquiescence in the face of hegemonic discourses. As Saul says, "Equilibrium [characterized by a balance between ideological stances and capacity for resistance to these ideologies], in the Western Experience, is dependent not just on criticism, but on non-conformism in the public place."[17] (Note, for example, such non-conforming actions introduced in Chapter 9.) Accordingly, we tend to look towards socio-constructive and transformative opportunities projected in Quadrant IV of our heuristic for more coherence between our educational and ethics aims. Here we anticipate more opportunity for critical reflections, active engagement with emerging ideas, and space for non-conformity – and hence opportunities for more authentic democratic engagement.

While we were aware that all quadrants of our heuristic would have some useful place in educational settings, the work of Mirian Vilela, Executive Director of the Earth Charter International Secretariat, had added depth and understanding to the possibilities. In her experiences, Quadrant I, for example, does not have to be simply a place for "Big Brother" authoritarians. Many of her students found working in this quadrant was a steppingstone towards reaching teaching opportunities in Quadrant IV. Through her insights, we have learned more about the work that can be done in Quadrants I, II, and III of our heuristic.

Still, there is ample literature to suggest that environmental ethics can meaningfully and fruitfully be conceived as something that continues to unfold – that needs continuous exploration. This is the work of Quadrant IV. For example, Aldo Leopold once wrote, "nothing so important as an ethic is ever 'written'."[18] He spoke of it as an on-going social evolution, and one that never stops. Implicit in these descriptions are participants who are constantly engaged in the reworking of relationships between themselves and, as he termed it then, the land. Implicit also, is an eschewing of any presumptions of a "true" or "final" ethic. Then he adds:

> Only the most superficial student of history supposes that Moses "wrote" the Decalogue; it evolved in the minds of a thinking

---

17   Ibid., 190.
18   Aldo Leopold, A *Sand County Almanac*, 263.

community, and Moses wrote a tentative summary of it for a "seminar." I say tentative because evolution never stops.[19]

Anthony Weston picks up this theme in a series of papers that come into focus in a paper titled, "Before environmental ethics".[20] Following Leopold, he argues that we have no idea where this field will take us. For Weston, environmental ethics is in need of a great deal of exploration. We should expect, at best, a long period of experimentation and uncertainty. So, it is clear that from within the field of environmental ethics (and before) there are strong voices promoting ethics as a process that produces, at best, tentative results and involves experimentation and uncertainty. Perhaps, following authors like Blenkinsop and Beeman[21] and Gruenewald,[22] researchers will find that documents like the Earth Charter can be a useful tool, alongside other narratives in environmental ethics, to enable this exploration – without circumscribing it.

Elaborating, Weston argues that our challenge is not to systematise environmental values, but rather to create the "space" for environmental values to evolve. By space, he is speaking about the social, psychological, and phenomenological preconditions that are needed to enable this evolutionary process. He is also speaking about the conceptual, experiential, and physical freedom to move and think. Here, Weston is concerned that individuals and groups *can* actually begin to create, or co-evolve, new values through everyday practices. Our job might thus be seen as "enabling environmental practice".[23]

Feminist scholars have also troubled the notion of universal codes of ethics. Classic works by writers such as Karen Warren[24] and Val Plumwood[25] argue that environmental ethics is not objective or disinterested, but rather, emerges as theory in process. According to these accounts, ethics need to be contextualised, narrated, and to hold a place for feelings and emotional understanding. A rich and growing body of feminist literature promises to lead researchers into rich new territory.

For now, we add one more line of critique that problematises ethics when conceived of as codes. Arne Næss[26] observes that human capacity for loving, based on moral duties or compliance with codes, is very limited. Often, in such circumstances, we act against our inclination but comply out of respect for the moral law. His alternative is to seek ways to do what

---

19  Ibid., 263.
20  Anthony Weston, "Before environmental ethics," 321-338.
21  See Sean Blenkinsop and Chris Beeman, "The Earth Charter, a radical document."
22  David Gruenewald, "A Foucauldian analysis of environmental education."
23  Anthony Weston, "Before environmental ethics."
24  Karen Warren, "The power and promise of ecological feminism," 125-146.
25  Val Plumwood, "Nature, self, and gender," 3-27; Val Plumwood, *Feminism and the mastery of nature*; Val Plumwood, *Environmental culture: The ecological crisis of reason.*
26  Arne Næss, "Self realization: An ecological approach to being in the world," 19-30; Arne Næss, *Life's Philosophy: Reason and Feeling in a deeper world.*

is right because of positive inclinations – out of joy, for example. When we do so, we perform a beautiful act. John Livingston[27] and Zygmunt Bauman[28] are similarly concerned about moral codes. For Livingston, they are unknown in nature and, as human creations, are more like prosthetic devices. For him it is important to develop an extended consciousness – beyond the mere self. Bauman argues that complying with moral codes reduces responsibilities for one's own moral actions; codes actively erode our moral impulses.

The examples in this section, reflecting Quadrant IV of our environmental ethics heuristic (Figure 11.3), point to conceptions of ethics more concerned with processes of engagement, experimentation, and transformation. They also encourage continued exploration of what ethics, and ethics research in environmental education, can become.[29] With these directions in mind, it would be inconsistent to remain fixed on any of the specific areas of ethics as, for example, those illustrated throughout this sourcebook or any specific statement of ethical principles.

However, we offer some caution. The heuristic presented here is designed to help practitioners clarify their conceptions of education and ethics, especially in the context of environmental education and, from there clarify their educational aims and pedagogical strategies. It is not meant to set standards for day-to-day practice. Many practitioners and researchers may ultimately aspire to the aims represented by one of these four quadrants, yet at times draw from the others in practice. As Vilela points out, work in these other quadrants can be steppingstones for gaining depth of understanding. At other times, we, as educators, can just be inconsistent. Again, we can look to Arne Næss who said, "to those who would like to be consistent: 'It's a high ideal to be consistent. And, you will achieve it when you die—not before.'"[30] Like Næss, we think the work we suggest is an on-going process, sometimes requiring small steps and reflexivity.

In the end, the heuristic presented here is just an analytical tool that can be used to critique current discourses, evaluate new initiatives, and find one's own place within present debates. We also see it as a tool that can illuminate tensions between explicit aims – in this case educational and participatory – and implicit messages in policies and practices. Finally, we encourage readers to adapt, develop, or reinvent this heuristic to suit their own needs and to aid in their own reflection on education and environmental ethics.

[27] John Livingston, *Rogue primate: An exploration of human domestication.*
[28] Zygmunt Bauman, *Postmodern ethics.*
[29] See for example Bob Jickling, "Making Ethics an Everyday Activity," 11-30; Bob Jickling, "Ethics research in environmental education," 20-34.
[30] Arne Næss and Bob Jickling, "Deep ecology and education," 58.

# References

Bauman, Zygmunt. *Postmodern ethics*. Oxford: Blackwell Publishers, 1993.

Blenkinsop, Sean and Beeman, Chris. "The Earth Charter, a radical document: A pedagogical response." *Factis Pax* 2, no. 1 (2008): 69-87.

Blühdorn, Ingolfur. "Political sociology and the cultural framing of environmental discourse: Depoliticisation, repoliticisation and the governance of unsustainability." Unpublished paper presented to Arts and Humanities Research Council network, Bath, U.K., 2010.

Earth Charter Initiative. (n.d.). "The Earth Charter initiative." Accessed November 4th, 2010. www.earthcharterinaction.org/content/

Globe and Mail, Advertisement. April 1, 2013, p. A14.

Gruenewald, David. "A Foucauldian analysis of environmental education: Toward the socio-ecological challenge of the Earth Charter." *Curriculum Inquiry* 34, no. 1 (2004): 71-107. https://doi.org/10.1111/j.1467-873X.2004.00281.x

Jickling, Bob. "Normalizing catastrophe: An educational response." *Environmental Education Research* 19, no. 2 (2013): 161-176. https://doi.org/10.1080/13504622.2012.721114

Jickling, Bob. "Ethics research in environmental education." *Southern African Journal of Environmental Education*, 22, (2005): 20-34. https://doi.org/10.1080/00958960309603496

Jickling, Bob. "Making Ethics an Everyday Activity: How Can We Reduce the Barriers?" *Canadian Journal of Environmental Education* 9, (2004): 11-30.

Jickling, Bob, Lotz-Sisitka, Heila, O'Donoghue, Rob, Ogbuigwe. Akpeizi. *Environmental Education, Ethics, and Action: A Workbook to Get Started*. Nairobi: UNEP, 2006.

Jickling, Bob, & Wals, Arjen E. J. "Globalization and environmental education: Looking beyond sustainability and sustainable development." *Journal of Curriculum Studies*, 40, no. 1 (2008): 1-21. https://doi.org/10.1080/00220270701684667

Jickling, Bob, & Wals, Arjen E. J. "Probing normative research in environmental education: Ideas about education and ethics." In *International Handbook of Research on Environmental Education*, ed. M. Brody, J. Dillon, R. B. Stevenson, and A.E.J. Wals. New York: Routledge, 2013: 74-86

Leopold, Aldo. A *Sand County Almanac: With essays on conservation from Round River*. New York: Ballantine, 1966. First published 1949/1953 by Oxford University Press.

Livingston, John. *Rogue primate: An exploration of human domestication*. Boulder, CO.: Roberts Rinehart Publishers, 1994.

Næss, Arne. "Self realization: An ecological approach to being in the world." In *Thinking like a mountain: Towards a council of all beings*, ed. J. Seed,

J. Macy, P. Fleming, and A. Næss. Gabriola Island, B.C.: New Society Publishers,1988: 19-30.

Næss, Arne. *Life's Philosophy: Reason and Feeling in a deeper world.* Athens, Georgia: University of Georgia Press, 2002.

Næss, Arne and Jickling, Bob. "Deep ecology and education: A conversation with Arne Næss." *Canadian Journal of Environmental Education* 5 (2000): 48-62.

Plumwood, Val. "Nature, self, and gender: Feminism, environmental philosophy, and the critique of reason." *Hypatia*, 6 no. 1 (1991): 3-27. https://doi.org/10.1111/j.1527-2001.1991.tb00206.x

Plumwood, Val. *Feminism and the mastery of nature.* New York: Routledge, 1993.

Plumwood, Val. *Environmental culture: The ecological crisis of reason.* London, UK: Routledge, 2002.

Saul, John Ralston. *The unconscious civilization.* Concord, Ontario: Anasi, 1995.

Sayer, Andrew. *Realism and social science.* London: Sage Publications, 2000. https://doi.org/10.4135/9781446218730

Van Harmelen, Ursula. "Education, ethics, and values: A response to Peter Blaze Corcoran's keynote address, Annual conference of the Environmental Education Association of Southern Africa 2003." *Southern African Journal of environmental education* 20 (2003): 124-128.

Weakland, Joseph P and Corcoran, Peter B. "The Earth Charter in higher education for sustainability." *Journal of Education for Sustainable Development* 3, no. 2 (2009): 151-158. https://doi.org/10.1177/097340820900300210

Warren, Karen. "The power and promise of ecological feminism." *Environmental Ethics* 12, no. 2 (1990): 125-146. https://doi.org/10.5840/enviroethics199012221

Weston, Anthony. "Before environmental ethics." *Environmental Ethics* 14, no. 4 (1992): 321-338. https://doi.org/10.5840/enviroethics19921444

# HANDOUT: THE EARTH CHARTER PRINCIPLES

## I. RESPECT AND CARE FOR THE COMMUNITY OF LIFE

1. Respect Earth and life in all its diversity.

   a. Recognize that all beings are interdependent and every form of life has value regardless of its worth to human beings.

   b. Affirm faith in the inherent dignity of all human beings and in the intellectual, artistic, ethical, and spiritual potential of humanity.

2. Care for the community of life with understanding, compassion, and love.

   a. Accept that with the right to own, manage, and use natural resources comes the duty to prevent environmental harm and to protect the rights of people.

   b. Affirm that with increased freedom, knowledge, and power comes increased responsibility to promote the common good.

3. Build democratic societies that are just, participatory, sustainable, and peaceful.

   a. Ensure that communities at all levels guarantee human rights and fundamental freedoms and provide everyone an opportunity to realize his or her full potential.

   b. Promote social and economic justice, enabling all to achieve a secure and meaningful livelihood that is ecologically responsible.

4. Secure Earth's bounty and beauty for present and future generations.

   a. Recognize that the freedom of action of each generation is qualified by the needs of future generations.

   b. Transmit to future generations values, traditions, and institutions that support the long-term flourishing of Earth's human and ecological communities.

## In order to fulfill these four broad commitments, it is necessary to:

## II. ECOLOGICAL INTEGRITY

5. Protect and restore the integrity of Earth's ecological systems, with special concern for biological diversity and the natural processes that sustain life.

   a. Adopt at all levels sustainable development plans and regulations that make environmental conservation and rehabilitation integral to all development initiatives.

   b. Establish and safeguard viable nature and biosphere reserves, including wild lands and marine areas, to protect Earth's life support systems, maintain biodiversity, and preserve our natural heritage.

c. Promote the recovery of endangered species and ecosystems.

d. Control and eradicate non-native or genetically modified organisms harmful to native species and the environment, and prevent introduction of such harmful organisms.

e. Manage the use of renewable resources such as water, soil, forest products, and marine life in ways that do not exceed rates of regeneration and that protect the health of ecosystems.

f. Manage the extraction and use of non-renewable resources such as minerals and fossil fuels in ways that minimize depletion and cause no serious environmental damage.

6. Prevent harm as the best method of environmental protection and, when knowledge is limited, apply a precautionary approach.

a. Take action to avoid the possibility of serious or irreversible environmental harm even when scientific knowledge is incomplete or inconclusive.

b. Place the burden of proof on those who argue that a proposed activity will not cause significant harm, and make the responsible parties liable for environmental harm.

c. Ensure that decision-making addresses the cumulative, long-term, indirect, long distance, and global consequences of human activities.

d. Prevent pollution of any part of the environment and allow no build-up of radioactive, toxic, or other hazardous substances.

e. Avoid military activities damaging to the environment.

7. Adopt patterns of production, consumption, and reproduction that safeguard Earth's regenerative capacities, human rights, and community well-being.

a. Reduce, reuse, and recycle the materials used in production and consumption systems, and ensure that residual waste can be assimilated by ecological systems.

b. Act with restraint and efficiency when using energy, and rely increasingly on renewable energy sources such as solar and wind.

c. Promote the development, adoption, and equitable transfer of environmentally sound technologies.

d. Internalize the full environmental and social costs of goods and services in the selling price, and enable consumers to identify products that meet the highest social and environmental standards.

e. Ensure universal access to health care that fosters reproductive health and responsible reproduction.

f. Adopt lifestyles that emphasize the quality of life and material sufficiency in a finite world.

8. Advance the study of ecological sustainability and promote the open exchange and wide application of the knowledge acquired.

   a. Support international scientific and technical cooperation on sustainability, with special attention to the needs of developing nations.

   b. Recognize and preserve the traditional knowledge and spiritual wisdom in all cultures that contribute to environmental protection and human well-being.

   c. Ensure that information of vital importance to human health and environmental protection, including genetic information, remains available in the public domain.

## III. SOCIAL AND ECONOMIC JUSTICE

9. Eradicate poverty as an ethical, social, and environmental imperative.

   a. Guarantee the right to potable water, clean air, food security, uncontaminated soil, shelter, and safe sanitation, allocating the national and international resources required.

   b. Empower every human being with the education and resources to secure a sustainable livelihood, and provide social security and safety nets for those who are unable to support themselves.

   c. Recognize the ignored, protect the vulnerable, serve those who suffer, and enable them to develop their capacities and to pursue their aspirations.

10. Ensure that economic activities and institutions at all levels promote human development in an equitable and sustainable manner.

    a. Promote the equitable distribution of wealth within nations and among nations.

    b. Enhance the intellectual, financial, technical, and social resources of developing nations, and relieve them of onerous international debt.

    c. Ensure that all trade supports sustainable resource use, environmental protection, and progressive labor standards.

    d. Require multinational corporations and international financial organizations to act transparently in the public good, and hold them accountable for the consequences of their activities.

11. Affirm gender equality and equity as prerequisites to sustainable development and ensure universal access to education, health care, and economic opportunity.

    a. Secure the human rights of women and girls and end all violence against them.

b. Promote the active participation of women in all aspects of economic, political, civil, social, and cultural life as full and equal partners, decision makers, leaders, and beneficiaries.

c. Strengthen families and ensure the safety and loving nurture of all family members.

12. Uphold the right of all, without discrimination, to a natural and social environment supportive of human dignity, bodily health, and spiritual well-being, with special attention to the rights of indigenous peoples and minorities.

a. Eliminate discrimination in all its forms, such as that based on race, color, sex, sexual orientation, religion, language, and national, ethnic or social origin.

b. Affirm the right of indigenous peoples to their spirituality, knowledge, lands and resources and to their related practice of sustainable livelihoods.

c. Honor and support the young people of our communities, enabling them to fulfill their essential role in creating sustainable societies.

d. Protect and restore outstanding places of cultural and spiritual significance.

## IV. DEMOCRACY, NONVIOLENCE, AND PEACE

13. Strengthen democratic institutions at all levels, and provide transparency and accountability in governance, inclusive participation in decision making, and access to justice.

a. Uphold the right of everyone to receive clear and timely information on environmental matters and all development plans and activities which are likely to affect them or in which they have an interest.

b. Support local, regional and global civil society, and promote the meaningful participation of all interested individuals and organizations in decision making.

c. Protect the rights to freedom of opinion, expression, peaceful assembly, association, and dissent.

d. Institute effective and efficient access to administrative and independent judicial procedures, including remedies and redress for environmental harm and the threat of such harm.

e. Eliminate corruption in all public and private institutions.

f. Strengthen local communities, enabling them to care for their environments, and assign environmental responsibilities to the levels of government where they can be carried out most effectively.

14. Integrate into formal education and life-long learning the knowledge, values, and skills needed for a sustainable way of life.

   a. Provide all, especially children and youth, with educational opportunities that empower them to contribute actively to sustainable development.

   b. Promote the contribution of the arts and humanities as well as the sciences in sustainability education.

   c. Enhance the role of the mass media in raising awareness of ecological and social challenges.

   d. Recognize the importance of moral and spiritual education for sustainable living.

15. Treat all living beings with respect and consideration.

   a. Prevent cruelty to animals kept in human societies and protect them from suffering.

   b. Protect wild animals from methods of hunting, trapping, and fishing that cause extreme, prolonged, or avoidable suffering.

   c. Avoid or eliminate to the full extent possible the taking or destruction of non-targeted species.

16. Promote a culture of tolerance, nonviolence, and peace.

   a. Encourage and support mutual understanding, solidarity, and cooperation among all peoples and within and among nations.

   b. Implement comprehensive strategies to prevent violent conflict and use collaborative problem solving to manage and resolve environmental conflicts and other disputes.

   c. Demilitarize national security systems to the level of a non-provocative defense posture, and convert military resources to peaceful purposes, including ecological restoration.

   d. Eliminate nuclear, biological, and toxic weapons and other weapons of mass destruction.

   e. Ensure that the use of orbital and outer space supports environmental protection and peace.

   f. Recognize that peace is the wholeness created by right relationships with oneself, other persons, other cultures, other life, Earth, and the larger whole of which all are a part.

# AFTERWORD

## CONTEXTUALISING ENVIRONMENTAL ETHICS IN THE CONTEMPORARY GLOBAL WORLD

In the last years during the production of this sourcebook (2017-2021), we have borne witness to an ever more worrying series of disastrous global climate and environmental events: Heat waves, forest fires, drought, floods, and hurricanes. In the same period, there have been significant instances of care for the more-than-human world that inspire hope. Hopeful global environmental action can be seen both in terms of direct care toward the more-than-human world, and also in the form of grassroots political action that aims to spur decision making that might facilitate more pro-environment policy on the part of global governments and other corporate institutions. It is notable, too, that such examples – of both hopeful and disastrous events – seem to reflect something about the nature of our times. As we learned from Andrew Sayer in Chapter 3, periods of major disruption and social change tend to prompt renewed interest in ethical debate.[1] We hope this sourcebook will respond, in part, to that trend.

In roughly alternating order of disastrous and hopeful, examples of these kinds of events include:

- In the 2017 Atlantic hurricane season, multiple category five storms – including hurricanes Harvey, Irma, and Maria – bared down on the south eastern United States, and many Caribbean Islands. This season caused billions of dollars of economic damage and more importantly immeasurable human suffering. Particularly hard hit was the island of Puerto Rico, a United States Territory. While mainland-American coastal areas experienced emergency response and restoration of critical infrastructure and services over the span of weeks and months, nearly one year following the landfall of hurricane Maria, Puerto Rico was still rebuilding.[2]

- In 2018, no elephants were poached within the North Luangwa National Park in Zambia and poaching in the region directly surrounding the park was reduced by 50%. This milestone represents a best-ever annual record for the park and region since this data has been tracked. Conservation authorities credit local community engagement in elephant conservation as the most significant factor in achieving this success.[3]

---

[1]   Andrew Sayer, *Realism and Social Science*.
[2]   See Umair Irfan, "Puerto Rico's Deadly Record Blackout Is Almost over."
[3]   John C. Cannon, "Community Buy-in Stamps out Elephant Poaching in Zambian Park."

- Monsoon rains led to catastrophic flooding in Bangladesh during August of 2017.[4] The intensity of flooding contributed to a phenomenon dubbed by some as "climate migrants" or "environmental refugees"[5] wherein climate related heavy-weather forces people to move. This often means migrating from rural to urban areas that, in turn, hastens urbanisation. This has a magnifying effect on sprawling cities that become even more susceptible to flooding, frequently as a result of poor infrastructure for managing the run off of torrential rain. Such forced migration patterns are rife with ethics quandaries arising from the social and ecological stresses created by many additional people arriving in cities in search of food, shelter, and jobs.[6]

- Throughout 2018 and 2019, Greta Thunberg, a teenage activist from Sweden, gained prominence for her role in spurring the "Fridays for Future" climate strike movement,[7] and again during the Global Climate Strike in September of 2019.[8] What began with Greta's solo protest outside the Swedish Parliament buildings quickly spread to Friday school strikes for climate justice undertaken by young people in Europe, and around the world. Between September 20 and 27, 2019, 7.6 million people of all ages participated in the largest ever activist mobilisation related to global climate change. The movement has attracted significant attention from media. While the actual impact of the event is yet unknown, its high profile may be a harbinger of growing political will to consider systemic changes in response to what many feel is a climate crisis.[9]

- New Zealand and India became the first countries in the world to assign human legal status to rivers in 2017. First, New Zealand's parliament voted to assign human legal equivalency to the Whanganui River, and to assign the responsibility for representing the river to a panel of two people, representing the Crown and Maori people, respectively.[10] Shortly thereafter, a court in India issued a ruling to recognise the environmental personhood of the Ganges and Yamuna Rivers; however, this ruling was later quashed on appeal by the country's Supreme Court. The higher court accepted arguments about the practical complications of environmental personhood such as holding the river accountable for drownings or flooding.[11] These split outcomes in New Zealand and India highlight the complexity of environmental personhood cases, but also the emergence of global

4    Wheeling, Kate. "How Climate Change Contributed to Massive Floods in South Asia."
5    John Vidal, "'Boats Pass over Where Our Land Was.'"
6    Ibid.
7    Fridays for Future, "About."
8    Global Climate Strike, "Global Climate Strike → A Historic Week."
9    Ibid.
10   BBC News, "NZ River given Legal Human Status."
11   BBC News, "Key India Rivers 'Not Living Entities.'"

political will for raising the ethical status of the more-than-human world.

- On March 11, 2020, the World Health Organization assessed that a novel coronavirus identified as COVID-19 could be considered a global pandemic.[12] In the weeks that followed, nonessential social, educational, and economic activities around the world slowed to a near halt as people responded to varying degrees of regional shelter-in-place orders. At the time of writing (January 2021), Johns Hopkins University COVID-19 Resource Centre calculates that nearly 100 million cases of COVID-19 and over two million related deaths had been documented globally across 191 countries and regions.[13] The pandemic has myriad connections to the issues of ethics that we discuss throughout the sourcebook,[14] but two points seem of primary importance for understanding the pandemic as both an environmental disaster and an obvious guidepost for environmental hope. First, is that COVID-19 has laid bare the inescapable link between human flourishing in relation to disease and the earth's natural environment. It has shown us that even the most well developed social, economic, or built structures cannot shield us from natural elements, because we interact with the environment at all times – in the context of a respiratory virus, literally with every breath. Second, swift human response to the pandemic has demonstrated our capacity to mount and implement massive agendas of change on a global scale when there is collective political and ethical will to do so. Responses to the pandemic executed during 2020 essentially neutralize longstanding social, economic, and logistical arguments against taking action to enable living in alignment with widely agreed upon environmental ethics. Simply put, COVID-19 has demonstrated the old cliché – *where there is a will, there is a way.*

We present these exemplars in the final pages of the sourcebook not for the purpose of extending our theorising in any new directions, but rather to point out the immediate and essential relevance of environmental ethics in the contemporary world. Throughout this book, we have explored issues of ethics that are deeply bound up with these specific recent events, and others like them. Ethics pervade both the causes and outcomes of the global pandemic, increasingly frequent "heavy weather" incidents, and socio-political activism together with their direct environmental actions that inspire hope for a liveable future on Earth. As we have made manifest throughout these pages, environmental issues require ongoing consideration, re-consideration, and action towards a more ethical world.

---

[12]  World Health Organization, "Timeline of WHO's Response to COVID-19."
[13]  Johns Hopkins Coronavirus Resource Center, "Home."
[14]  Indeed, we debated writing an additional chapter to include a deeper dive into the environmental ethics of COVID-19, but opted not to delay publication in order to allow time for additional writing.

Education can be at the heart of these decisions, and educators have significant opportunities for fostering a knowledge of ethics – and indeed, ethical knowledge related to the environment. In writing this book, we have aimed to provide resources and inspiration for educators to achieve maximum leverage of these opportunities.

In this Afterword, we point to examples of worldwide environmental challenges and parallel movements of hope to emphasize the significance of environmental ethics – and therefore education. Educators can and must prepare people for more frequent and more nuanced application of ethical lenses to environmental thinking and actions: in terms of direct strategies that mitigate environmental degradation, and in broader socio-political movements designed to draw attention to environmental crises and to catalyse political will.

Throughout this sourcebook, we have offered educators practical exercises that are connected to theoretical insights. They have been crafted to encourage learners to think about ethics and the environment across a range of formal and informal learning contexts. In Chapter 1, we introduced these reflective pedagogical strategies through a sensitising graphic that indicated three emergent themes in the development of this sourcebook: Pedagogies that situate ethics-led learning in history and context, pedagogies that explore values and the moral impulse, and pedagogies that stimulate and guide ethical action. We now conclude with a summary of these overarching pedagogies that have been woven throughout the preceding chapters. The bullet points in the following sections are offered as key takeaways and are linked to chapters throughout the book. They provide link-back opportunities for readers who may want to review specific aspects of the three pedagogical themes. Readers may notice some repetition of concepts or ideas across the three themes; we have maintained these multiplicities in order to demonstrate their inherent interrelationships.

## Pedagogies that situate ethics-led learning in history and context

- Ethics-led learning is a pedagogical ethos in which teaching and learning move forth following an ethical pathway. This means that education follows ethics as they develop, rather than launching off some rigid and pre-ordained ethical foundation that education comes to orbit at a distance. (Chapter 1)
- What is ethical is not fixed or universally preset, but nor is it entirely relative. Ethics are negotiated temporally, culturally, regionally, and in many other ways. Ethics-led teaching and learning encourages habits of mind and behavioural etiquettes that respond to this

dynamic landscape of living in good relationships. (Chapter 1 and Chapter 6)

- Through a relational lens, ethics may be viewed as dynamic, but not completely free floating; when ethics are conceptualised as a state of living that is "being *for* the Other",[15] then living ethically becomes anchored to an underlying human inclination or impulse to be in good relation with one another, more-than-human living things, and other life-supporting elements that make up our environments. (Chapter 8)

- While our attention may often be called to environmental ethics at times of crisis or other pre-set occasions – Earth day, Earth hour, Global Climate Strike, etc. – an ethics-led learning approach aims to shift attention to the big and small ways that environmentally ethical living pervades peoples' everyday lives. (Chapter 1)

- To forward a concept of ethics that is important in everyday living, we have suggested the value of stories as vehicles for taking up ethical work; stories can come in the form of oral traditions, literature, and even contemporary multi-media content. The stories we tell impact our lives and how we live them, sometimes in obvious ways but more often in subtle and inconspicuous ways. (Chapter 2)

- If stories are ethical touchstones, and ethics are best understood as a part of our everyday lives, then the work of *doing* ethics must include investigating and even interrogating the stories that we use to define ourselves individually and collectively. What do our stories say about us and how we live on our shared planet? Are these the stories that we need? How can we change our stories if they are not helping us to live ethically in our environments? (Chapter 3)

## Pedagogies that explore values and the moral impulse

- We present Jim Cheney and Anthony Weston's[16] work as building on the second bullet point above. Following their work, we propose that it is in the nature of knowledge to be entwined with ethical practice – hence an ethics-based epistemology. This to say that their work suggests that everyday actions inform our sense of knowing the world and "right" ways to live in it. Cheney and Weston describe these careful daily actions as *etiquette*, or an ethics-led path to knowledge. (Chapter 7)

- Etiquettes of daily living can unfold in all arenas of life – home life, social life, professional life, etc. Studying these actions to better understand our intentions and their impacts is a foundational aspect

---

15 Zygmunt Bauman. *Postmodern Ethics*, 13.
16 Jim Cheney and Anthony Weston, "Environmental Ethics as Environmental Etiquette," 115-34.

of developing and maintaining theories of knowledge that can help human communities to grapple with the socio-environmental challenges that they face. (Chapter 7)

■ When values come into conflict, they can create ethical quandaries. We describe quandaries as complex value conflicts where either/or solutions are not satisfactory in untangling the conflict. In facing a quandary, we propose that having a practice of reflecting on our daily actions – or etiquettes – can be helpful in pointing forward directions. (Chapter 4)

■ Ethics do not begin and/or end with humans, but rather transect the human and more-than-human world. Some ethicists (Singer, Regan, Taylor) argue for a duty to extend ethical consideration to the more-than-human to the degree that beings display *telos*, or agency, purpose of being. These ethical framings can be helpful to the degree that they enable ethics language and help us tell our own ethics stories. However, we are also interested in conceptions that attend to moral impulses as a basis for developing ethical relationships with the more-than-human world (Livingston, Bauman). (Chapter 4)

■ Developing strong moral impulses toward the more-than-human world is largely dependent on having regular and repeated experiences in nature. The Scandinavian notion of *friluftsliv* (free air life) is one cultural conception that encourages people to be outdoors. Throughout the sourcebook, we offer activities that can encourage educators to take their participants outdoors to have experiences that may help them to develop practices of ethical action toward the natural world. Such experiences and ethical actions may give way to more locally developed cultural understandings akin to *frilufsliv*. This is an example of the importance of reciprocal cycles of experiences in natural environments and ethical reflection. (Chapter 5)

## Pedagogies that stimulate and guide ethical action

■ Ethics and environmental ethics are not only intellectual exercises, but are deeply rooted in practice and lived experience. Effective teaching and learning about environmental ethics connect intellectual and emotional aspects of ethics with actions intended to bring ethical thinking and feeling into reality. (Chapter 8)

■ The notion of ethics as a relational practice is furthered through Bauman's proximity thesis.[17] He argues that the capacity or likelihood for ethical consideration of others lays largely in the proximity an individual has with the Other. This proximity may be conceptual, as in, viewing the other as morally adjacent as an agent that is not "me," but is a moral extension of me. But, proximity may also be

---

[17] See Zygmunt Bauman, *Postmodern Ethics*.

significantly physical. Proximity with another increases the degree to which individuals or groups may extend moral consideration. In other words, the closer we are, the stronger an ethical impulse we are likely to feel for one another. Problematically, many aspects of contemporary social practice and habits of daily living threaten both physical and metaphysical proximity and erode the ethical impulse that can be fostered through closeness. (Chapter 8)

- While critique – the naming of problems and resisting ethical injustices – are significant aspects of bringing ethics into action, these elements do not tell the whole story. Anthony Weston's work helps us to recognise that ethical action is not simply about problematising, but importantly includes what Weston calls reimagining. For Weston, reimagining the world means seeing beyond problems in order to embrace many innovative, and sometimes unusual, solutions. At its most effective, reimagining can involve turning problems upside down, and inside out in order to view situations anew, and begin to create a more ethical world. (Chapter 9)

- Instances of the kind of reimagining exercises that Weston calls for may be found in the ideas and actions promoted by many contemporary social movements. Liberation movements like Occupy, Black Lives Matter, Idle No More, The Women's March, and Fridays for Future offer opportunities for exploring reimagined futures through acts that help reveal social structures that are often hidden in plain sight. They can also highlight joints and fault-lines of those structures that could be redesigned in order to bring about a reimagined world. (Chapter 9)

- As was laid out in the theme grouping on history and context, ethics is neither a free-for-all of relativity, nor is it universally fixed. Still, it is important to be able to make decisions about how to act in ways that are good or right in terms of how we inhabit our planet together. To decide how to act in the face of a plurality of possible "right" actions, Bhaskar asks us to seek "grounds for preferring one belief … to another".[18] This deliberative process is the hard work of doing ethics, and requires a suite of thinking tools that allow for considerations about intention, probable outcomes, history, and social context in order to commit to actions that are socially and environmentally just. (Chapter 10)

- We populate our thinking tool kit with Bhaskar's assemblage of eight thinking tools, expressed here as four elemental questions: What is, and what is not "right"? What could be different? What should be, considering a holistic vision of self and others? What realistic, concrete action can we take? We offer Bhaskar's action heuristic as a helpful starting point for deciding how to act at times when

---

[18] Roy Bhaskar, *Possibility of naturalism*, 58.

ethical uncertainty clouds decision making. We do not presume that Bhaskar's elements are the only way to approach ethical problem solving. Indeed, other models certainly exist, and we hope that through the process of dancing with ethical dilemmas that teachers and learners can develop new processes for thinking and acting ethically. (Chapter 10)

## Wrapping up ... for now

We began these last words with contemporary exemplars of disaster incidents **and** movements of hope related to environmental issues. We highlighted the ethical dimensions of these points of hope and despair. We did this as a way to emphasise the significance of practicing environmental ethics as a way to enable a more harmonious existence of humans alongside more-than-human others on earth. We then use this sense of both urgency and complexity as a springboard from which to contextualise the three pedagogical themes developed throughout the book. We believe they have potential for improving the status of environmental ethics through education.

The three pedagogical themes – pedagogies that situate ethics-led learning in history and context, pedagogies that explore values and the moral impulse, and pedagogies that stimulate and guide ethical action – are bound up in the examples of hopeful and despairing incidents that we have offered above. Such stories cultivate dynamic reactions of despair, anxiety, desperation on one hand; and hope, courage, and resilience on the other. These feelings are manifestations of the complexity of environmental ethics in action.

Throughout the sourcebook, we have aspired to highlight the complexity implicit in ethics work. However, it is important to emphasise that complexity must not be an excuse for delay on critical environmental decisions. So often in our times, naysayers on environmental issues – most notably climate change – use complexity as a scapegoat for inaction that ultimately protects the status quo. In the face of such stall tactics, we latch on to the third pedagogical theme – the stimulation and guidance of ethical action. This crucial theme reminds us that we must appreciate the complexity of environmental issues, and also move to act in environmentally ethical ways if we hope to avert catastrophic planetary outcomes. This guiding and stimulating action theme further reminds us that the global changes that are needed to redirect the earth from climate collapse are not intrinsic to the kinds of high-profile movements and events that we profiled at the outset of this chapter. Those events do serve as effective stories for spurring hope, but we suggest that each of those lighthouse examples – assigning environmental personhood to rivers, a week of 7.8 million global climate strikers – is fundamentally comprised of myriad *everyday ethics*.[19]

Indeed, we contend that it is meaningful engagement with environmental ethics in everyday kinds of ways by individuals and local communities that can pay dividends toward mitigating large scale events and stimulating policy developments like the ones outlined at the beginning of this chapter. Those events and policy decisions, it is important to remember, are planned and executed by people who have been touched by educators. Educators have a tremendous capacity to spark inspiration, motivation, and impetus for enacting environmental ethics to generations of people young and old who are hungry to act for a more environmentally sustainable world. We wrote this book for just those educators, and hope it finds its way into their hands, hearts, and programme plans.

## References

Bauman, Zygmunt. *Postmodern Ethics*. Oxford: Blackwell Publishers, 1993.

BBC News. "Key India Rivers 'Not Living Entities'," July 7, 2017, https://www.bbc.com/news/world-asia-india-40537701

BBC News. "NZ River given Legal Human Status," March 15, 2017, sec. Asia. https://www.bbc.com/news/world-asia-39282918

Bhaskar, Roy. *Possibility of Naturalism*. 3rd ed. London: Routledge, 1998.

Cannon, John C. "Community Buy-in Stamps out Elephant Poaching in Zambian Park." Mongabay Environmental News, April 22, 2019. https://news.mongabay.com/2019/04/community-buy-in-stamps-out-elephant-poaching-in-zambian-park/.

Cheney, Jim and Anthony Weston. "Environmental Ethics as Environmental Etiquette: Toward an Ethics-Based Epistemology." *Environmental Ethics* 21, no. 2 (1999): 115-34. https://doi.org/10.5840/enviroethics199921226

Fridays for Future. "About," accessed October 2, 2019, https://www.fridaysforfuture.org/about

Global Climate Strike. "Global Climate Strike → A Historic Week," accessed September 29, 2019, https://globalclimatestrike.net.

Irfan, Umair. "Puerto Rico's Deadly Record Blackout Is Almost over," *Vox*, July 3, 2018, https://www.vox.com/2018/7/3/17530814/puerto-rico-power-blackout-over-hurricane-maria

Jickling, Bob. "Making Ethics an Everyday Activity: How Can We Reduce the Barriers?" *Canadian Journal of Environmental Education* 9 (2004): 11-26.

Johns Hopkins Coronavirus Resource Center. "Home," accessed January 20, 2021, https://coronavirus.jhu.edu/

Sayer, Andrew. *Realism and Social Science*. London: SAGE, 2000. https://doi.org/10.4135/9781446218730

---

[19] Bob Jickling, "Making Ethics an Everyday Activity," 11-26.

Vidal, John. "'Boats Pass over Where Our Land Was': Climate Refugees in Bangladesh | John Vidal," *The Guardian*, January 4, 2018, https://www.theguardian.com/global-development/2018/jan/04/bangladesh-climate-refugees-john-vidal-photo-essay

Wheeling, Kate. "How Climate Change Contributed to Massive Floods in South Asia," *Pacific Standard*, accessed October 2, 2019, https://psmag.com/environment/how-climate-change-contributed-to-massive-floods-in-south-asia

World Health Organization. "Timeline of WHO's Response to COVID-19," accessed January 20, 2021, https://www.who.int/emergencies/diseases/novel-coronavirus-2019/interactive-timeline#!

# INDEX

Note: Locators in *italics* refer to figures and photographs.

www.ingramcontent.com/pod-product-compliance
Lightning Source LLC
Chambersburg PA
CBHW080130270326
41926CB00021B/4427